Books published in this series focus on the theory and computational foundations, advanced methodologies and practical applications of machine learning, ideally combining mathematically rigorous treatments of a contemporary topics in machine learning with specific illustrations in relevant algorithm designs and demonstrations in real-world applications. The intended readership includes research students and researchers in computer science, computer engineering, electrical engineering, data science, and related areas seeking a convenient medium to track the progresses made in the foundations, methodologies, and applications of machine learning.

Topics considered include all areas of machine learning, including but not limited to:

- Decision tree
- Artificial neural networks
- Kernel learning
- Bayesian learning
- Ensemble methods
- Dimension reduction and metric learning
- Reinforcement learning
- Meta learning and learning to learn
- Imitation learning
- Computational learning theory
- Probabilistic graphical models
- Transfer learning
- Multi-view and multi-task learning
- Agents and Multi-Agent Systems
- Graph neural networks
- Generative adversarial networks
- Federated learning
- Large Language Models
- Multimodal Learning
- Transformer/Diffusion Models
- Generative Artificial Intelligence
- Bio-inspired Learning Models
- Embodied AI
- Explainable AI and Ethics

This series includes monographs, introductory and advanced textbooks, and state-of-the-art collections. Furthermore, it supports Open Access publication mode.

Yanyan Shen · Lei Chen
Editors

Graph Neural Network Training

From Data Management Perspective

Springer

Editors
Yanyan Shen
School of Computer Science
Shanghai Jiao Tong University
Shanghai, China

Lei Chen
Hong Kong University of Science
and Technology (Guangzhou)
Guangzhou, China

ISSN 2730-9908 ISSN 2730-9916 (electronic)
Machine Learning: Foundations, Methodologies, and Applications
ISBN 978-981-95-5794-3 ISBN 978-981-95-5795-0 (eBook)
https://doi.org/10.1007/978-981-95-5795-0

This Springer imprint is published by the registered company Springer Nature Singapore Pte Ltd.
The registered company address is: 152 Beach Road, #21-01/04 Gateway East, Singapore 189721, Singapore

Preface

Graph Neural Networks (GNNs) have revolutionized the extraction of insights from complex and interconnected data in recent years. These models have proven essential across both industry and academia, powering applications ranging from recommendation systems to drug discovery. However, as graphs expand to billions of nodes and trillions of edges, training costs present a significant barrier. The recursive message passing inherent to GNNs creates substantial overhead in communication, data movement, and hardware utilization. This efficiency bottleneck impedes research advancement and constrains real-world deployment, prompting the authors to develop this book. Writing the book required synthesizing knowledge across the evolving domains of machine learning, systems, and database research. The authors organized diverse innovative ideas into a cohesive narrative by examining GNN training through a data management lens, unifying techniques that span the complete graph data processing pipeline from preprocessing through model training.

This book serves multiple audiences, including graduate students exploring scalable graph learning, machine learning engineers deploying GNNs on massive datasets, and system developers applying data management principles to GNN training workloads. The authors summarize the latest progress in accelerating GNN training, encompassing sampling algorithms, data transfer methods, novel data layouts, and hardware-aware optimizations. The content spans both centralized and distributed settings while addressing challenges across static and dynamic graphs. The objective is to equip readers with an understanding of the current landscape and empower them to advance the field. By mastering these techniques, readers will be prepared to build next-generation scalable graph learning systems that unlock the full potential of GNNs.

Shanghai, China
Guangzhou, China
December 2025

Yanyan Shen
Lei Chen

Contents

Chapter 1
Introduction to Graph Neural Network Training on Large Graphs

Yanyan Shen and Lei Chen

Abstract Graph Neural Networks (GNNs) are a powerful paradigm for modeling graph-structured data, but their reliance on iterative neighborhood aggregation makes training computationally expensive, especially on large graphs. To address the resulting scalability challenges, the database community has recently focused on optimizing GNN training from a data management perspective. This chapter summarizes the end-to-end GNN training pipeline along the graph data lifecycle, encompassing the graph preprocessing, the batch generation, the data transfer, and the model training stages, and reviews representative techniques that improve efficiency at both stage level and end-to-end pipeline level. We distinguish the challenges of static and dynamic graph settings, discussing efficient training strategies for each. Finally, we outline how the remainder of this book is organized around scalable GNN training methodologies.

1.1 Background

Graph Neural Networks (GNNs) have emerged as a powerful tool for extracting complex structural and semantic information from massive graphs. They have gained continuous popularity in a broad range of graph-based applications including social network analysis [15], financial risk assessment [2], knowledge graph construction [34], recommendation systems [10], etc.

Despite the variation in model architectures, most GNNs adopt the message passing scheme that exchanges information among nodes based on the graph structure.

Y. Shen (✉)
Shanghai Jiao Tong University, Shanghai, China
e-mail: shenyy@sjtu.edu.cn

L. Chen (✉)
The Hong Kong University of Science and Technology, Hong Kong SAR, China

The Hong Kong University of Science and Technology (Guangzhou), Guangzhou, China
e-mail: leichen@cse.ust.hk

Y. Shen and L. Chen (eds.), *Graph Neural Network Training*, Machine Learning: Foundations, Methodologies, and Applications,
https://doi.org/10.1007/978-981-95-5795-0_1

This scheme involves recursive neighbor information aggregation, which poses significant challenges when training GNNs on large graph datasets. These challenges include (1) high communication cost among machines for full-batch GNN training, (2) excessive data movement between CPU and GPU for mini-batch GNN training, and (3) low hardware utilization during model training. As highlighted by Sancus [27], when performing full-batch training of the GCN model [20] on the Ogbn-Products [14] dataset with an eight-GPU machine, the communication time accounts for over 90% of the end-to-end training time, even for highly optimized GNN training systems. Similarly, data movement dominates the end-to-end mini-batch GNN training time. The situation worsens for larger graph datasets like Ogbn-Papers100M [14]. Moreover, a recent study [1] has revealed that better computation resource utilization can lead to approximately 10× speedup in per-epoch GNN training efficiency.

As graph data continues to grow in size, developing efficient GNN training systems becomes increasingly crucial. At a high level, current training systems process graph data on hybrid CPU-GPU platforms, following the intricate computational logic of forward and backward propagation and facilitating data movement across heterogeneous hardware. With a longstanding history of studying efficient methods for complex graph processing, database researchers have recently directed their efforts toward exploring data management techniques to accelerate the GNN training process. This line of research generally progresses in parallel with GNN model development but emphasizes achieving better training efficiency without compromising model performance.

In this chapter, we provide an overview of the recent efforts for accelerating GNN training on massive graphs. Based on the graph data lifecycle, we view the training process through the lens of four major stages (Fig. 1.1): graph preprocessing, batch generation, data transfer, and model training. We discuss innovative data management ideas at different stages that improve GNN training efficiency. Both centralized and distributed training environments are considered. Due to the distinct characteristics of static and dynamic graphs and their respective GNN models, we first focus on efficient GNN training on large static graphs. Subsequently, we delve into the efforts aimed at training GNNs on large dynamic graphs.

1.2 Graph Preprocessing

To enhance the efficiency of GNN training, existing works propose to perform necessary graph preprocessing. In this section, we introduce three common preprocessing techniques, namely graph size reduction, graph partitioning, and graph reordering, and discuss how they influence the later batching and transferring efficiency of the GNN training.

Graph Size Reduction. Training on large graphs can be time-consuming, and a straightforward idea to address this problem is to compact the training graph while maintaining the model accuracy. To do this, three common techniques are available:

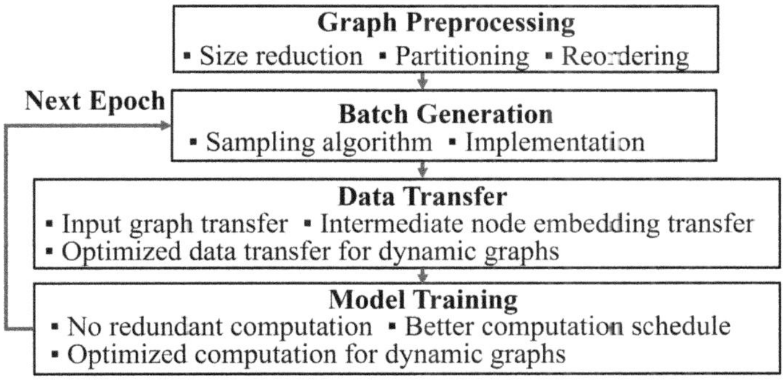

Fig. 1.1 GNN training workflow

graph coarsening, graph sparsification, and graph condensation. Graph coarsening [24] tries to group nodes in the original graph into super-nodes so that the graph size is reduced. Graph sparsification [19] removes the edges in the original graph to eliminate redundant and insignificant relationships. Graph condensation is different from the above two techniques in that it tries to construct synthetic smaller graphs. The graph condensation problem is often formulated as a bi-level optimization problem: we update the synthetic data in the outer loop based on the neural network trained on the synthetic data in the inner loop. Solving the bi-level optimization problem can be computationally expensive or even intractable. To improve the graph condensation efficiency, existing works try to (1) speed up the inner loop [6], or (2) avoid the bi-level optimization by solving the inner loop model training problem with a closed-form solution [30]. There are also works about graph-free condensation for better generalization ability [35].

Graph Partitioning. Due to the limited device memory, graph partitioning on graph structure and feature data is employed to create smaller subgraphs for distributed GNN training. The current partitioning algorithms aim to reduce the communication cost and balance the workloads across different subgraphs. To partition the graph structure, we can use traditional graph partitioning algorithms [4, 27] (e.g., METIS partitioning, random partitioning) or strategies tailored for GNN training [17, 26]. For feature data, P^3 [9] optimize the communication cost by vertically partitioning.

Graph Reordering. Reordering the vertices in graph is an effective way to enhance locality, which benefits both model computation and data transfer in GNN training. For example, GNNAdvisor [29] reorders the graph selectively to improve the locality of computation in graph convolution kernels. DUCATI [33] also proposes to reorder the graph based on access frequency to facilitate a lightweight cache construction and lookup for the topology data of the graph dataset. When traversing the graph topology with the unified memory, the HALO reordering [13] can improve the utilization of PCIe bandwidth and accelerate batch data transfer.

1.3 Batch Generation

In GNN training, graph sampling has emerged as a powerful technique for generating batches. It selects nodes and their associated edges from the original graph. The sampled subgraphs serve as the batches that are fed into the downstream GNN training.

Graph Sampling Algorithms. A variety of graph sampling approaches have been tailored specifically to the characteristics of GNN training workflows. Node-wise sampling aggregates messages from a subset of neighbors, mitigating the computational challenges arising from the exponentially increasing dependencies in GNNs. Layer-wise sampling employs independent sampling for each network layer and leverages importance sampling to maintain a constant sample size, effectively controlling the computational growth. Subgraph-wise sampling generates a subgraph and reuses it for the computation across all layers. Furthermore, certain studies have highlighted the benefits of delicately considering the spatial locality in feature storage and extraction, which can significantly reduce the transfer cost. Feature-oriented sampling [32] initiates the sampling process by selecting node features, which then guides the formation of the corresponding subgraph for a mini-batch.

Graph Sampling Implementations. Efficient sampling involves leveraging the computational capabilities of available hardware resources in the system. The GPU serves as an ideal hardware choice for this purpose due to its excellent parallel computing power. Meanwhile, generating the sampled subgraphs on the GPU avoids the additional costs of transferring graph samples into the device memory. Nextdoor [16] is a pioneer work that performs graph sampling on a single GPU. It loads the entire graph into device memory and proposes a transit-parallelism strategy to accelerate the sampling process by aggregating the same transit nodes. To enhance scalability through the extensive capacity of host memory, the Unified Virtual Addressing (UVA) technique is adapted for graph sampling [4, 33]. It stores the graph structure in host memory and executes graph sampling on GPUs, taking advantage of the combined resources of both the host and the device.

1.4 Data Transfer

Efficient data transfer between CPU and GPU(s), or among multiple GPUs, is paramount for ensuring high GNN training efficiency on large-scale graphs. This is because the required data for model computation may not be locally available. The specific data that needs to be transferred depends on the applied GNN training approach, i.e., mini-batch-based or full-batch-based GNN training.

Transferring Input Graph Data. For mini-batch-based GNN training on large-scale graphs, cross-device transfer of the input graph data (initial node feature vectors and graph structure information) is a necessity. However, this data transfer process incurs substantial costs during training. To alleviate this issue, a common practice is to utilize local graph data caches to reduce the volume of data being transferred and employ sophisticated cache management policies. PaGraph [26] and GNNLab [31] exploit node feature locality, caching node feature vectors locally prior to training with certain hotness metrics. DUCATI [33] further exploits the locality of graph structure data and introduces an additional graph structure data GPU cache. It also leverages a refined allocator for managing both types of caches, to minimize the aggregate transfer volume of both types of graph input data between CPUs and GPUs.

Transferring Intermediate Node Embeddings. For full-batch-based GNN training, a large graph is typically partitioned across devices. In this setup, intermediate node embeddings, capturing the neighborhood-aggregated node representations across different GNN layers, need to be transferred across devices to enable full-neighbor aggregation of target nodes. DGCL [3] finds optimal transfer routes for every node's intermediate embeddings given the system topology to minimizing the corresponding transfer costs. Neutronstar [28] achieves reduced transfer cost for intermediate node embeddings at the expense of redundant bottom-up neighborhood aggregation locally. Sancus [27] proposes a skip-broadcast data transfer mechanism that flexibly bypasses intermediate node embedding transfer between GPUs during specific training epochs. This is guided by several refined management regarding the staleness of the embeddings.

1.5 Model Training

The efficiency of model training (i.e., the forward and backward propagation) is also important for the overall training efficiency. Researchers have explored approaches to avoid redundant computation [18] or optimize the computation schedule, i.e., optimizing resource allocation of different operations for better resource utilization in a collaborative CPU-GPU computation setup [25], optimizing training parallelism [25], and improving the efficiency of computation graphs and operators on hardware [7, 8].

1.6 Training on Dynamic Graphs

Apart from the traditional static graphs, many real-world graphs exhibit dynamic behaviors with evolving interaction patterns. Typically, there are two types of dynamic graphs: discrete-time dynamic graphs (DTDGs) and continuous-time dynamic graphs (CTDGs), which are typically handled by dynamic GNNs (D-GNNs) and temporal GNNs (T-GNNs), respectively. The dynamic nature of these graphs introduces new challenges and optimizations for accelerating data transfer and computation during GNN training.

Optimized Data Transfer for Dynamic Graphs. For D-GNN training, Chakaravarthy et al. [5] capitalize the topological similarities across different graph snapshots, and design a graph-difference-based transfer method to reduce the transfer cost of graph snapshots. For T-GNN training, ETC [11] identifies the redundant data access patterns and utilizes a redundancy-aware data access policy to reduce the transfer volume of input features. SIMPLE [12] further captures the complex entanglement effect between time and frequency information in CTDGs, and reduces the data transfer cost via dynamic data placement on GPU.

Optimized Computation for Dynamic Graphs. Li et al. [21] identify the reusable intermediate embeddings across graph snapshots in D-GNN training, and propose cache policies that maximize the saved model computation time for the next snapshot. Orca [22] also caches intermediate embedding in T-GNN training. It employs a reuse-distance-based cache policy, which optimally minimizes the overall re-computation costs under the given GPU cache budget. Zebra [23] points out the temporal effect of node influence in T-GNN training. It performs single-layer aggregation over the most influential neighbors, improving computational efficiency.

1.7 Outline of This Book

The remainder of this book is structured as follows. Chapter 2 introduces foundational concepts and training techniques related to Graph Neural Networks, covering static and dynamic graph settings, as well as various training hardware scenarios (single-GPU, multi-GPU, distributed clusters). Chapter 3 uncovered two data transfer issues that hinder the efficient and scalable training of GNNs and proposes a scalable graph sampling algorithm and an efficient GNN training system to address them. Chapter 4 focuses on the techniques to accelerate the computation for scalable GNN training and proposes two solutions from the model execution planning aspect and the sparse operator optimization aspect, respectively. Chapter 5 studies the distributed GNN training problem and proposes a staleness-aware communication-avoiding decentralized GNN system. Chapter 6 explores the area of temporal Graph Neural Networks training, including two complementary system prototypes to address the bottlenecks of ill-suited batching and high input data loading costs on large-scale graphs.

References

1. Xin Ai, Qiange Wang, Chunyu Cao, Yanfeng Zhang, Chaoyi Chen, Hao Yuan, Yu Gu, and Ge Yu. 2024. NeutronOrch: Rethinking Sample-based GNN Training under CPU-GPU Heterogeneous Environments. *Proc. VLDB Endow.* 17, 8 (2024), 1995–2008. https://doi.org/10.14778/3659437.3659453
2. Wendong Bi, Bingbing Xu, Xiaoqian Sun, Zidong Wang, Huawei Shen, and Xueqi Cheng. 2022. Company-as-tribe: Company financial risk assessment on tribe-style graph with hierarchical

graph neural networks. In *KDD*. 2712–2720.

3. Zhenkun Cai, Xiao Yan, Yidi Wu, Kaihao Ma, James Cheng, and Fan Yu. 2021. DGCL: An efficient communication library for distributed GNN training. In *EuroSys*. 130–144.
4. Zhenkun Cai, Qihui Zhou, Xiao Yan, Da Zheng, Xiang Song, Chenguang Zheng, James Cheng, and George Karypis. 2023. DSP: Efficient GNN training with multiple GPUs. In *PPoPP*. 392–404.
5. Venkatesan T Chakaravarthy, Shivmaran S Pandian, Saurabh Raje, Yogish Sabharwal, Toyotaro Suzumura, and Shashanka Ubaru. 2021. Efficient scaling of dynamic graph neural networks. In *SC*. 1–15.
6. Junfeng Fang, Xinglin Li, Yongduo Sui, Yuan Gao, Guibin Zhang, Kun Wang, Xiang Wang, and Xiangnan He. 2024a. EXGC: Bridging Efficiency and Explainability in Graph Condensation. In *WWW*. 721–732. https://doi.org/10.1145/3589334.3645551
7. Jingzhi Fang, Yanyan Shen, Yue Wang, and Lei Chen. 2020. Optimizing DNN computation graph using graph substitutions. *VLDB* 13, 12 (2020), 2734–2746.
8. Jingzhi Fang, Yanyan Shen, Yue Wang, and Lei Chen. 2024b. STile: Searching Hybrid Sparse Formats for Sparse Deep Learning Operators Automatically. *Proceedings of the ACM on Management of Data* 2, 1 (2024), 1–26.
9. Swapnil Gandhi and Anand Padmanabha Iyer. 2021. P3: Distributed deep graph learning at scale. In *OSDI*. 551–568.
10. Chen Gao, Xiang Wang, Xiangnan He, and Yong Li. 2022. Graph neural networks for recommender system. In *WSDM*. 1623–1625.
11. Shihong Gao, Yiming Li, Yanyan Shen, Yingxia Shao, and Lei Chen. 2024a. ETC: Efficient Training of Temporal Graph Neural Networks over Large-scale Dynamic Graphs. *VLDB* 17, 5 (2024), 1060–1072.
12. Shihong Gao, Yiming Li, Xin Zhang, Yanyan Shen, Yingxia Shao, and Lei Chen. 2024b. SIMPLE: Efficient Temporal Graph Neural Network Training at Scale with Dynamic Data Placement. *Proceedings of the ACM on Management of Data* 2, 3 (2024), 1–25.
13. Prasun Gera, Hyojong Kim, Piyush Sao, Hyesoon Kim, and David Bader. 2020. Traversing large graphs on GPUs with unified memory. *VLDB* 13, 7 (2020), 1119–1133.
14. Weihua Hu, Matthias Fey, Marinka Zitnik, Yuxiao Dong, Hongyu Ren, Bowen Liu, Michele Catasta, and Jure Leskovec. 2020. Open graph benchmark: Datasets for machine learning on graphs. *NeurIPS* 33 (2020), 22118–22133.
15. Lokesh Jain, Rahul Katarya, and Shelly Sachdeva. 2023. Opinion leaders for information diffusion using graph neural network in online social networks. *TWEB* 17, 2 (2023), 1–37.
16. Abhinav Jangda, Sandeep Polisetty, Arjun Guha, and Marco Serafini. 2021. Accelerating graph sampling for graph machine learning using GPUs. In *EuroSys*. 311–326.
17. Zhihao Jia, Sina Lin, Mingyu Gao, Matei Zaharia, and Alex Aiken. 2020a. Improving the accuracy, scalability, and performance of graph neural networks with roc. *MLSys* 2 (2020), 187–198.
18. Zhihao Jia, Sina Lin, Rex Ying, Jiaxuan You, Jure Leskovec, and Alex Aiken. 2020b. Redundancy-free computation for graph neural networks. In *KDD*. 997–1005.
19. Xunqiang Jiang, Tianrui Jia, Yuan Fang, Chuan Shi, Zhe Lin, and Hui Wang. 2021. Pre-training on large-scale heterogeneous graph. In *KDD*. 756–766.
20. Thomas N. Kipf and Max Welling. 2017. Semi-Supervised Classification with Graph Convolutional Networks. In *ICLR*. https://openreview.net/forum?id=SJU4ayYgl
21. Haoyang Li and Lei Chen. 2021. Cache-based GNN system for dynamic graphs. In *CIKM*. 937–946.
22. Yiming Li, Yanyan Shen, Lei Chen, and Mingxuan Yuan. 2023a. Orca: Scalable Temporal Graph Neural Network Training with Theoretical Guarantees. *Proc. ACM Manag. Data* 1, 1 (2023), 52:1–52:27. https://doi.org/10.1145/3588737
23. Yiming Li, Yanyan Shen, Lei Chen, and Mingxuan Yuan. 2023b. Zebra: When Temporal Graph Neural Networks Meet Temporal Personalized PageRank. *VLDB* 16, 6 (2023), 1332–1345.
24. Zhiyuan Li, Xun Jian, Yue Wang, and Lei Chen. 2022. CC-GNN: A community and contraction-based graph neural network. In *ICDM*. IEEE, 231–240.

25. Zhiyuan Li, Xun Jian, Yue Wang, Yingxia Shao, and Lei Chen. 2024. DAHA: Accelerating GNN Training with Data and Hardware Aware Execution Planning. *VLDB* 17, 6 (2024), 1364–1376. https://www.vldb.org/pvldb/vol17/p1364-li.pdf
26. Zhiqi Lin, Cheng Li, Youshan Miao, Yunxin Liu, and Yinlong Xu. 2020. PaGraph: Scaling GNN training on large graphs via computation-aware caching. In *SoCC*. 401–415.
27. Jingshu Peng, Zhao Chen, Yingxia Shao, Yanyan Shen, Lei Chen, and Jiannong Cao. 2022. SANCUS: Staleness-Aware Communication-Avoiding Full-Graph Decentralized Training in Large-Scale Graph Neural Networks. *VLDB* 15, 9 (2022), 1937–1950. https://doi.org/10.14778/3538598.3538614
28. Qiange Wang, Yanfeng Zhang, Hao Wang, Chaoyi Chen, Xiaodong Zhang, and Ge Yu. 2022. NeutronStar: Distributed GNN Training with Hybrid Dependency Management. In *SIGMOD*. 1301–1315. https://doi.org/10.1145/3514221.3526134
29. Yuke Wang, Boyuan Feng, Gushu Li, Shuangchen Li, Lei Deng, Yuan Xie, and Yufei Ding. 2021. GNNAdvisor: An adaptive and efficient runtime system for GNN acceleration on GPUs. In *OSDI*. 515–531.
30. Zhe Xu, Yuzhong Chen, Menghai Pan, Huiyuan Chen, Mahashweta Das, Hao Yang, and Hanghang Tong. 2023. Kernel Ridge Regression-Based Graph Dataset Distillation. In *KDD*. 2850–2861.
31. Jianbang Yang, Dahai Tang, Xiaoniu Song, Lei Wang, Qiang Yin, Rong Chen, Wenyuan Yu, and Jingren Zhou. 2022. GNNLab: a factored system for sample-based GNN training over GPUs. In *EuroSys*. 417–434.
32. Xin Zhang, Yanyan Shen, and Lei Chen. 2022. Feature-Oriented Sampling for Fast and Scalable GNN Training. In *ICDM*. IEEE, 723–732.
33. Xin Zhang, Yanyan Shen, Yingxia Shao, and Lei Chen. 2023. DUCATI: A Dual-Cache Training System for Graph Neural Networks on Giant Graphs with the GPU. *Proc. ACM Manag. Data* 1, 2 (2023), 166:1–166:24. https://doi.org/10.1145/3589311
34. Yongqi Zhang, Quanming Yao, Yingxia Shao, and Lei Chen. 2019. NSCaching: simple and efficient negative sampling for knowledge graph embedding. In *ICDE*. IEEE, 614–625.
35. Xin Zheng, Miao Zhang, Chunyang Chen, Quoc Viet Hung Nguyen, Xingquan Zhu, and Shirui Pan. 2024. Structure-free graph condensation: From large-scale graphs to condensed graph-free data. *NeurIPS* 36 (2024).

Chapter 2
Preliminaries

Shihong Gao and Xin Zhang

Abstract This chapter provides foundational concepts for Graph Neural Networks, covering static and dynamic graph settings. For static graphs, we define core components such as graph structure, message passing mechanisms, and training processes. We also discuss common graph analytics tasks, datasets, and sparse storage formats. For dynamic graphs, we introduce continuous-time dynamic graphs, temporal GNNs, and their training pipelines, emphasizing chronological constraints and hybrid CPU-GPU data layouts. Finally, we review hardware-aware training scenarios and system optimizations to reduce data transfer and improve bandwidth utilization. This section serves as a comprehensive primer for scalable GNN training across diverse settings.

2.1 Common Definitions

Graph Neural Networks (GNNs) are a class of deep learning models designed for learning on graph-structured data. We first present the key definitions:

- **Graph**. A graph is defined as a tuple $G = (V, E)$, where $V = \{v_1, v_2, \ldots, v_N\}$ denotes the set of N nodes, and $E = \{e_1, e_2, \ldots, e_M\}$ denotes the set of M edges connecting pairs of nodes, typically with $e = (v_i, v_j)$ representing an edge from node v_i to node v_j. Graphs can be directed or undirected and are often equivalently represented using an adjacency matrix.
- **Node/Edge Features**. Each node or edge can be associated with an attribute vector to describe its characteristics. Node features are usually represented as a matrix $X_V \in \mathbb{R}^{N \times d}$, where each node has a d-dimensional feature vector, and edge features are represented as $X_E \in \mathbb{R}^{M \times d}$. For instance, in social networks, node features may represent user profiles, while edge features may encode communication frequency or relationship types.

S. Gao (✉) · X. Zhang
The Hong Kong University of Science and Technology, Hong Kong SAR, China
e-mail: sgaoar@connect.ust.hk

X. Zhang
e-mail: xzhanggb@connect.ust.hk

Y. Shen and L. Chen (eds.), *Graph Neural Network Training*, Machine Learning: Foundations, Methodologies, and Applications,
https://doi.org/10.1007/978-981-95-5795-0_2

- **Message Passing Mechanism**. The core computational process of GNNs typically involves a message passing (or neighborhood aggregation) mechanism. Each target node aggregates messages (i.e., features) from its neighbors and updates its own representation using a learnable aggregation function. Formally, in a two-layer GNN, the feature update at node v can be approximated as:

$$h_v^{(k)} = \sigma\left(W^{(k)} \cdot \mathrm{AGG}\left(\{h_u^{(k-1)} : u \in \mathcal{N}(v)\}\right) + b^{(k)}\right), \tag{2.1}$$

where $\mathcal{N}(v)$ denotes the set of neighbors of node v, AGG is the aggregation function (e.g., mean, weighted sum, attention mechanism), σ is a nonlinear activation function, and $W^{(k)}$, $b^{(k)}$ are learnable parameters at the k-th layer.
- **Training Process and Computation Graph**. The training of GNNs is analogous to that of traditional neural networks, including forward propagation, loss computation, and gradient-based backpropagation. A key difference lies in the dynamic construction of computation graphs per iteration. For each target node, a computation subgraph is formed by retrieving its K-hop neighbors, akin to unfolding the graph into a tree. Multi-layer aggregation is then performed on this subgraph, followed by loss calculation and parameter updates.
- **Sampling**. Real-world graphs are often large-scale, making it impractical to process the entire graph per batch. To address this, training systems typically adopt sampling strategies: from a set of target nodes, a fixed number of neighbors are sampled (either randomly or via importance) to construct subgraphs for training. Sampling significantly reduces per-batch computation and memory usage.
- **Neighbor Aggregation**. This refers to the process by which a target node collects information from its neighbors, typically involving a linear transformation followed by summation, averaging, or attention-weighted aggregation. For example, GCN [1] uses symmetric normalization of the adjacency matrix for convolution, while GAT [2] assigns learnable attention weights to neighbors. The choice of aggregation function directly impacts the model's expressiveness and computational efficiency.
- **Memory Management**. GNN training is memory-intensive, especially on large graphs with high-degree nodes. Common memory optimization techniques include layer-wise sampling or subgraph sampling to reduce per-iteration memory load, as well as asynchronous I/O and memory caching strategies to reuse intermediate results. These will be discussed in detail for single-GPU and multi-GPU scenarios in subsequent sections.

2.2 Graph Neural Networks on Static Graphs

2.2.1 *Full Graph Training*

We define a graph structure as $\mathcal{G} = (\mathcal{V}, \mathcal{E})$, comprising a set of N nodes $\mathcal{V} = \{v_1, \ldots, v_N\}$ and a corresponding set of edges $\mathcal{E} \subseteq \mathcal{V} \times \mathcal{V}$. The topological connections are represented by a symmetric adjacency matrix $\mathbf{A}$, where each entry $A_{ij} = 1$ indicates an edge between nodes v_i and v_j, and $A_{ij} = 0$ otherwise. To facilitate effective message passing, widely used models like GCN [3] utilize a symmetrically normalized adjacency matrix $\hat{\mathbf{A}} = \bar{\mathbf{D}}^{-\frac{1}{2}}\bar{\mathbf{A}}\bar{\mathbf{D}}^{-\frac{1}{2}}$. Here, $\bar{\mathbf{A}} = \mathbf{A} + \mathbf{I}_N$ incorporates self-loops for each node, and $\bar{\mathbf{D}} \in \mathbb{R}^{N \times N}$ represents the diagonal degree matrix derived from $\bar{\mathbf{A}}$.

From a matrix operations perspective, the feature propagation at the ℓ-th layer of a GNN [4, 5] can be generalized as:

$$\mathbf{H}^{(\ell)} = \sigma\left(\mathbf{H}^{(\ell-1)}, \hat{\mathbf{A}}; \mathbf{W}^{(\ell-1)}\right). \tag{2.2}$$

In this formulation, $\mathbf{H}^{(0)} = \mathbf{X} \in \mathbb{R}^{N \times F}$ serves as the initial input, where each row corresponds to the feature vector of node v_i. The term $\mathbf{W} \in \mathbb{R}^{F \times F}$ denotes the learnable weight matrix, and σ represents a nonlinear activation function (e.g., ReLU). Note that we omit the superscript $^{(\ell)}$ when the context implies a specific layer. The training process involves two distinct phases:

- **Forward Phase.** During the forward pass, the model aggregates neighbor information and applies a linear transformation. This computational flow is mathematically expressed as:

$$\mathbf{Z}^{(\ell)} = \hat{\mathbf{A}}\mathbf{H}^{(\ell-1)}\mathbf{W}^{(\ell-1)}, \tag{2.3}$$

$$\mathbf{H}^{(\ell)} = \sigma\left(\mathbf{Z}^{(\ell)}\right). \tag{2.4}$$

- **Backward Phase.** The gradients are computed recursively via the chain rule to update the model parameters. Let $\delta^{(\ell)} = \nabla_{\mathbf{Z}^{(\ell)}}\mathcal{L}$ be the error term with respect to the intermediate representation $\mathbf{Z}^{(\ell)}$ given the loss function $\mathcal{L}$. The recurrence relation for error propagation is derived as:

$$\delta^{(\ell-1)} = \frac{\partial \mathcal{L}}{\partial \mathbf{Z}^{(\ell-1)}} = \delta^{(\ell)}\hat{\mathbf{A}}\left(\mathbf{W}^{(\ell-1)}\right)^{\mathsf{T}} \odot \sigma'\left(\mathbf{Z}^{(\ell-1)}\right), \tag{2.5}$$

where $\sigma'(\cdot)$ indicates the derivative of the activation function. Consequently, the gradient for the weight matrix $\mathbf{W}^{(\ell-1)}$ is calculated by:

$$\nabla_{\mathbf{W}^{(\ell-1)}}\mathcal{L} = \frac{\partial \mathcal{L}}{\partial \mathbf{W}^{(\ell-1)}} = \delta^{(\ell)}\hat{\mathbf{A}}\left(\mathbf{H}^{(\ell-1)}\right)^{\mathsf{T}}. \tag{2.6}$$

A complete training *epoch* comprises these forward and backward passes, concluding with the parameter optimization step using a learning rate η:

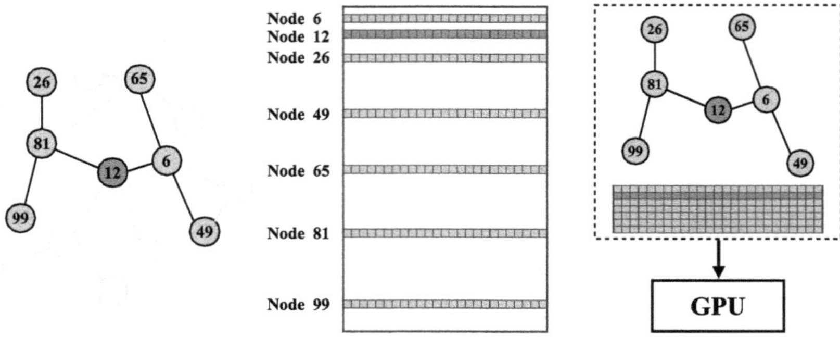

Fig. 2.1 Preparation of one mini-batch given an input graph and the corresponding node features

$$\mathbf{W}^{(\ell-1)} \leftarrow \mathbf{W}^{(\ell-1)} - \eta \nabla_{\mathbf{W}^{(\ell-1)}} \mathcal{L}. \tag{2.7}$$

The training iterates until model convergence is achieved. While the formulation above adopts a matrix-centric view typical for full-batch training, modern approaches often interpret this process through the lens of message passing [6, 7], particularly when dealing with sophisticated GNN architectures or sampling-based training strategies.

2.2.2 Mini-Batch Training

The inherent scalability limitations of full-batch training on massive graphs led to the adoption of mini-batch strategies, pioneered by frameworks like GraphSAGE [8]. Rather than processing the entire graph topology simultaneously, this approach operates on sampled subgraphs. For each iteration, a mini-batch is constructed to provide the local structural context and feature information required for the model to generate predictions for a specific set of target nodes.

We illustrate the mechanics of this process in Fig. 2.1, using a two-layer GraphSAGE model with a neighbor sampling fanout of {2, 2}. Taking node 12 as the target, the generation of a mini-batch typically involves two distinct phases.

- **Adj-Sampling (Structure Sampling)**. The system recursively traverses the graph structure. Starting from the target node 12, it accesses the adjacency list to randomly select neighbors (e.g., nodes 6 and 81), and repeats this process for the subsequent layer. This phase is characterized by random, fine-grained memory accesses as the sampler chases pointers through the graph topology.
- **Nfeat-Selecting (Feature Extraction)**. Once the computation subgraph is determined, the system collects the indices of all involved nodes. It then executes a bulk retrieval operation to fetch the corresponding feature vectors. In contrast to

the scattered access pattern of graph sampling, feature extraction typically allows for more compact and high-throughput memory operations.

Upon completion, the sampled structure and the gathered features are packaged into a mini-batch. The model then proceeds to compute the loss for node 12, execute backpropagation, and update the weights.

In standard implementation paradigms, the CPU is responsible for the data-intensive preparation tasks, *Adj-Sampling* and *Nfeat-Selecting*, storing the results in host memory. Subsequently, the prepared mini-batches are transferred over the PCIe bus to the GPU for the compute-intensive model training.

Beyond the node-wise neighbor sampling employed by GraphSAGE, the research community has developed a spectrum of sampling methodologies. These can be broadly categorized into node-wise, layer-wise, and subgraph-wise sampling. In the following sections, we will delve into the algorithmic details of these categories, reviewing representative works and analyzing the trade-offs they present regarding sampling overhead and training convergence.

Node-Wise Sampling

Node-wise sampling focuses on sampling a subset of neighbors for each node during the aggregation process. This approach reduces the computational cost by limiting the number of neighbors considered at each layer.

For each node, a fixed number of neighbors, which are usually referred to as fanouts, are randomly sampled at each layer. Then, the aggregation is performed only over the sampled neighbors, reducing the computation. The representative works of node-wise sampling are GraphSAGE [8] and PinSage [9].

GraphSAGE introduces a general framework for inductive representation learning on graphs. It employs node-wise sampling to sample a fixed number of neighbors for each node, enabling scalable training on large graphs. PinSage is an extension of GraphSAGE. PinSage uses importance-based sampling to select neighbors, prioritizing nodes with higher influence.

These works are simple to implement and computationally efficient since they reduce memory and computation by limiting the number of neighbors. They are suitable for large-scale graphs with high-degree nodes. However, the sampling process may lose important structural information if critical neighbors are not sampled. In addition, the node-wise sampling can lead to high variance in gradients, affecting convergence of GNN training. To make things worse, the node-wise sampling leads to exponential growth of total batch size with respect to the model depth. Consider the GraphSAGE sampling with 10 as fanout at each layer, the total batch size is $O(10^k)$, where k is the number of GNN layers. Such exponential growth significantly limits the scalability of node-wise sampling to deep GNN models.

Layer-Wise Sampling

Layer-wise sampling differs from node-wise sampling by posing sampling limitations to each layer independently. This approach ensures that the number of sampled nodes are consistent across layers, avoiding the exponential growth with respect to the model depths. The sampling process is depicted as follows. At each layer, a subset of nodes is sampled independently considering factors such as variance, connectivity, and contribution to loss. The aggregation is performed only over the paths connecting sampled nodes across different layers, ensuring consistency of training. The representative works of layer-wise sampling are FastGCN [10] and LADIES [11].

FastGCN treats node sampling at each layer as an importance sampling problem, where nodes are sampled based on their contribution to the loss function. It samples nodes independently at each layer, reducing the computational cost. However, the connectivity of sampled nodes across layers are usually sacrificed. LADIES improves upon FastGCN by sampling nodes based on the adjacency matrix, ensuring that the sampled nodes are connected and preserving graph structure.

The advantages of layer-wise sampling include (1) reducing redundancy by sampling nodes independently at each layer and (2) enhancing efficiency compared to node-wise sampling for deep GNNs by avoiding the neighbor explosion. However, layer-wise sampling may still lose structural information if critical nodes are not sampled. And importance sampling can introduce additional computational overhead, which may even surpass the actual GNN training computation, as pointed out by some works [12, 13].

Subgraph-Wise Sampling

Subgraph-wise sampling involves sampling entire subgraphs from the original graph and performing computations only on the sampled subgraphs. This approach is particularly effective for mini-batch training. The detail of sampling process is presented as follows. Given the original graph, we extract only a subgraph from it, typically by selecting a subset of nodes and their associated edges. The GNN is trained on the sampled subgraph instead of the original graph, reducing the computational cost. Representatives of this sampling approach include ClusterGCN [14] and GraphSAINT [12].

ClusterGCN first partitions all nodes of original graph into many densely connected clusters using graph clustering algorithms. Then, at each training step, one or more clusters are randomly chosen and combined to form a subgraph, which is used as the mini-batch for GNN training. This approach ensures that the sampled subgraphs are dense and representative of the original graph. However, the graph partitioning/clustering algorithm leveraged by ClusterGCN is usually computationally expensive on large and dense graphs. The preprocessing time of ClusterGCN, which is majorly spent on graph partitioning/clustering, can exceed the time of later model training. To address the problem, GraphSAINT proposes to induce subgraphs from sampled nodes or edges. GraphSAINT proposes and compares several sampling approaches to form the nodes/edges that are used for subgraph induction. To ensure that the sampled subgraphs are unbiased w.r.t. sampling frequency and preserve the graph structure, GraphSAINT introduces the normalization and variance reduction techniques.

The strength of subgraph-wise sampling is the capability to preserves local structure within the sampled subgraphs. The independent nature among sampled graphs makes it suitable for mini-batch training and distributed computing. It also reduces memory usage by working on smaller subgraphs. However, its weaknesses lie in the potentially expensive graph clustering and partitioning computations. In addition, subgraph-wise sampling may introduce bias if the sampling process does not preserve the global structure.

Here we present the overall comparison of the three mainstream sampling style and related methods. In summary, node-wise, layer-wise, and subgraph-wise sampling are effective strategies for scaling GNNs to large graphs. All these methods aim to reduce the computational cost of GNN training by sampling a subset of the graph. The node-wise sampling limits a fixed number neighbors for each target node at each layer, but suffers from the "neighborhood explosion" problem for deep GNN models. The layer-wise sampling addresses the "neighborhood explosion" problem by limiting the total number of sampled nodes at each layer, but encounters high sampling costs and certain lost of structure information. Subgraph-wise sampling preserves more local structure information and can avoid the expensive layer-wise sampling cost as well as the neihgbor explosion problem.

From the granularity perspective, node-wise sampling operates at the node level, layer-wise sampling at the layer level, and subgraph-wise sampling at the graph level. From the sampling overhead perspective, node-wise sampling has the lowest overhead, while layer-wise and subgraph-wise sampling may require additional preprocessing overload. The choice of method depends on the specific application, graph size, and desired trade-offs between efficiency and accuracy.

2.2.3 Common Tasks and Datasets for Graph Neural Networks

GNNs have been successfully applied to a wide range of tasks, including node classification, link prediction, and graph classification. Each of these tasks addresses different challenges in graph analysis and leverages unique datasets for benchmarking and evaluation. Below, we introduce these tasks and discuss commonly used datasets for each.

First, node classification is a fundamental task in graph analysis, where the goal is to predict the label or category of nodes in a graph based on their features and the graph structure. This task is widely used in applications such as social network analysis, recommendation systems, and bioinformatics. Commonly used datasets include Cora [15], Citeseer [15], PubMed [15], and Reddit [14]. Cora is citation network dataset consisting of 2,708 scientific publications (nodes) and 5429 citation links (edges). Each node has a 1433-dimensional binary feature vector representing the presence of specific words, and the task is to classify each publication into one of seven categories. PubMed is a biomedical literature dataset with 19,717 scientific publications (nodes) and 44,338 citation links (edges). Each node has a 500-dimensional feature vector derived from TF-IDF weights of words, and the task is to classify publications into three categories. Citeseer is a citation network dataset with 3327 publications (nodes) and 4732 citation links (edges). Each node has a 3703-dimensional feature vector, and the task is to classify publications into six categories. Reddit is a social network dataset with 232,965 posts (nodes) and 11,606,919 edges representing interactions. Each node has a 602-dimensional feature vector, and the task is to classify posts into one of 41 communities.

Second, link prediction aims to predict the existence of edges (links) between pairs of nodes in a graph. This task is crucial for applications such as recommender systems, knowledge graph completion, and social network analysis. Commonly used datasets include Facebook [16], Protein–Protein Interaction (PPI) [17], BlogCatalog [18], and UCI Message [19]. Facebook is a social network dataset with 4039 users (nodes) and 88,234 friendships (edges). The task is to predict missing friendships between users. PPI is a biological network dataset with 3852 proteins (nodes) and 37,841 interactions (edges). The task is to predict potential interactions between proteins. BlogCatalog is a social network dataset with 10,312 bloggers (nodes) and 333,983 friendships (edges). The task is to predict missing social connections. UCI Message is a communication network dataset with 1899 users (nodes) and 20,296 messages (edges). The task is to predict future interactions between users.

Third, graph classification involves predicting the label or property of an entire graph. This task is essential for applications such as molecular property prediction, social network analysis, and chemical compound classification. Commonly used datasets for this task includes MUTAG [20], PROTEINS[21], NCI1 [22], and IMDB-BINARY [23]. MUTAG is a dataset of 188 chemical compounds (graphs) with 7 discrete labels. Each graph represents a molecule, and the task is to classify molecules based on their mutagenic effect on bacteria. PROTEINS is a dataset of 1113 protein structures (graphs) with 2 labels. Each graph represents a protein, and the task is to classify proteins as enzymes or non-enzymes. NCI1 is a dataset of 4,110 chemical compounds (graphs) with 2 labels. Each graph represents a molecule, and the task is to classify molecules based on their activity against non-small cell lung cancer. IMDB-BINARY is a dataset of 1,000 movie collaboration graphs with 2 labels. Each graph represents an ego network of actors, and the task is to classify the genre of the movie (action or comedy).

2.2.4 Sparse Graph Representation and Storage

A standard graph dataset is composed of two primary data structures: the adjacency matrix representing topology and the feature matrix representing node attributes.

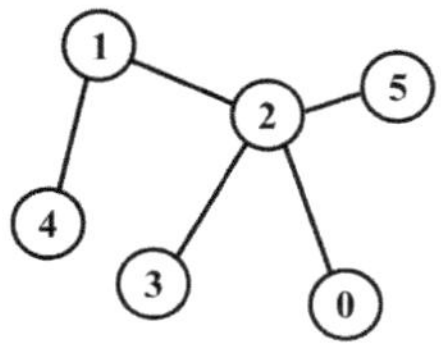

$$\begin{bmatrix} 0 & 0 & 1 & 0 & 0 & 0 \\ 0 & 0 & 1 & 0 & 1 & 0 \\ 1 & 1 & 0 & 1 & 0 & 1 \\ 0 & 0 & 1 & 0 & 0 & 0 \\ 0 & 1 & 0 & 0 & 0 & 0 \\ 0 & 0 & 1 & 0 & 0 & 0 \end{bmatrix}$$

$$\begin{aligned} \text{col\_index} &= [0, 1, 3, 7, 8, 9, 10] \\ \text{row\_index} &= [2, 2, 4, 0, 1, 3, 5, 2, 1, 2] \\ \text{values} &= [1, 1, 1, 1, 1, 1, 1, 1, 1, 1] \end{aligned}$$

Topology **Adjacency Matrix** **Compressed Sparse Columns**

Fig. 2.2 Adjacency matrix stored in CSC format

While node features are typically dense and can be efficiently managed as a continuous 2-D tensor of shape (N, f) where N denotes the node count and f the feature dimensionality, the topological data requires more specialized handling.

Real-world graphs exhibit significant sparsity, making dense matrix storage impractical. Consequently, compressed formats such as Coordinate List (COO), Compressed Sparse Row (CSR), and Compressed Sparse Column (CSC) are standard in graph processing. In the context of GNN training, particularly for sampling-based frameworks like GraphSAGE [8], the Compressed Sparse Column (CSC) format has become the de facto standard adopted by systems such as PyG [24] and DGL [25].

Figure 2.2 illustrates the mechanics of the CSC format, which compacts the adjacency matrix into three linear arrays:

- **`col_ptr`** (or `col_index`): This array serves as an index map. For any given node v_i (where $0 \leq i < N$), the range $[\texttt{col\_ptr}[i], \texttt{col\_ptr}[i+1])$ defines the segment in the subsequent arrays that corresponds to node v_i's connections.
- **`row_idx`** (or `row_index`): This array stores the row indices of non-zero elements, effectively listing the source nodes for incoming edges.
- **`values`**: This optional array holds the edge weights or attributes associated with the connections.

The prevalence of CSC in GNN systems is driven by the algorithmic requirements of the message-passing phase. Since GNNs typically aggregate information from a node's incoming neighbors (in-neighbors), the CSC layout ensures that all in-neighbors of a target node are stored contiguously in memory. This spatial locality is critical for performance, enabling the development of highly optimized C++ and CUDA sampling kernels that minimize random memory access and maximize bandwidth utilization.

2.3 Graph Neural Networks on Dynamic Graphs

This section establishes the theoretical foundations for learning on evolving graph structures. We formally define Continuous-Time Dynamic Graphs (CTDGs), introduce the architecture of Temporal Graph Neural Networks (T-GNNs), and delineate the standard pipeline employed for training these models on large-scale datasets.

2.3.1 Continuous-Time Dynamic Graphs

Definition 1 (Continuous-Time Dynamic Graph (CTDG)) A CTDG is formalized as a stream of time-stamped interaction events, denoted by $\mathcal{G} = \{\alpha(t_1), \alpha(t_2), \ldots\}$, ordered such that $t_1 \leq t_2 \leq \ldots$. Each interaction event is encapsulated by a tuple $\alpha(t) = (v_i, v_j, e_{ij}(t), t)$. Here, v_i and v_j represent the interacting source and destination nodes, respectively. t denotes the precise timestamp of the interaction, and $e_{ij}(t)$ is a feature vector encoding the attributes of this specific temporal edge.

Structurally, a CTDG functions as a temporal multigraph, allowing for multiple concurrent or sequential edges between any node pair (v_i, v_j), distinguished by their timestamps and feature attributes. Our focus on CTDGs, as opposed to Discrete-Time Dynamic Graphs (DTDGs), is driven by their superior fidelity in modeling real-world phenomena. Unlike DTDGs, which aggregate events into coarse-grained snapshots, CTDGs preserve the exact timing of interactions. This granularity provides a more expressive framework for capturing irregular temporal patterns and the continuous evolution inherent in dynamic systems [26–30].

A CTDG captures a rich spectrum of temporal events beyond simple interaction additions. These include event deletions and event updates, which are reflected through changes in the associated descriptive features [31]. In addition, node-level events—such as node additions, deletions, or feature updates—can be naturally represented as self-interaction events (i.e., edges where the source and target are the same node). As such, these events can be uniformly defined using the same temporal event formulation introduced in Definition 1.

2.3.2 Temporal Graph Neural Networks

Temporal Graph Neural Networks (T-GNNs) have demonstrated strong capabilities in representation learning over dynamic graphs. Existing state-of-the-art T-GNNs for Continuous-Time Dynamic Graphs (CTDGs) [27, 28, 30] largely follow a unified architectural paradigm that consists of two key components: node state update and temporal message passing.

Node State Update. Since nodes often have interaction histories of varying lengths, neighborhood sampling alone may be insufficient for capturing long-term temporal dependencies. To address this, most T-GNN models[27, 29, 30, 32] maintain a dedicated memory module that stores a state vector for each node, encoding its historical interaction information. Specifically, for a node v_i, a state vector $s_i(t)$ summarizes its temporal context up to time t. Upon the arrival of a new interaction $\alpha(t) = (v_i, v_j, e_{ij}(t), t)$, the model updates the states of both v_i and v_j as follows:

$$s_i(t) = \textbf{UPDATE}(s_i(t^-), s_j(t^-), e_{ij}(t), \phi(t - t^-)), \tag{2.8}$$

$$s_j(t) = \textbf{UPDATE}(s_j(t^-), s_i(t^-), e_{ij}(t), \phi(t - t^-)), \tag{2.9}$$

where **UPDATE** is typically implemented using RNN or GRU modules, and $\phi(\cdot)$ denotes a time encoding function that transforms the elapsed time $\Delta t = t - t^-$ into a vector representation.

Temporal Message Passing. When a new interaction $\alpha(t) = (v_i, v_j, e_{ij}(t), t)$ occurs, the model generates a temporal embedding $h_i^\ell(t)$ for node v_i (and similarly for v_j) through the following procedure:

$$N_i^\ell(t) = \textbf{SAMPLE}(\mathcal{G}, v_i, t), \tag{2.10}$$

$$h_i^\ell(t) = \textbf{AGGREGATE}\left(\left\{h_j^{\ell-1}(t^-) \,||\, e_{ij}(t^-) \,||\, \phi(t - t^-) \mid (v_j, t^-) \in N_i^\ell(t)\right\}\right), \quad \forall \ell = 1, \ldots, L, \tag{2.11}$$

$$h_j^0(t^-) = s_j(t^-) + \textbf{MLP}(x_j). \tag{2.12}$$

Here, **SAMPLE** performs temporal neighborhood sampling, selecting historical interactions involving v_i prior to timestamp t. Notably, temporal sampling accounts for timestamps, allowing the same node to be sampled multiple times at different moments. A common strategy is *top-k recent neighbor sampling* [27, 29, 32], which retrieves the k most recent neighbors of v_i based on interaction times. The **AGGREGATE** function, often implemented via attention mechanisms, aggregates the temporal information to produce the embedding at each layer. At the first layer ($\ell = 0$), the input features consist of the node's current state vector and its static feature x_j (if available). If initial node features are absent, only the state vectors are used.

2.3.3 *Training of Temporal Graph Neural Networks*

Generally, offline training of Temporal Graph Neural Networks (T-GNNs) is performed in chronological order. Below, we detail the data layout, preprocessing steps, and key training stages involved in training T-GNNs on large-scale dynamic graphs.

Data Layout. Prior studies [26–30, 32] often train T-GNNs using an *all-on-GPU* approach, where both the storage of input data and the training computations rely

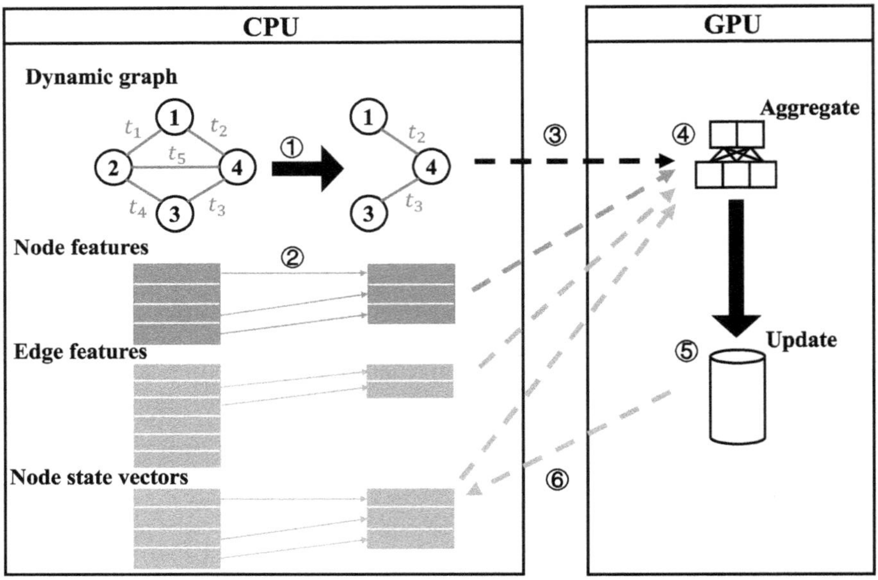

Fig. 2.3 Overview of the hybrid CPU-GPU data management strategy for T-GNN training. To accommodate large-scale dynamic graphs, the system retains the full graph topology, feature matrices, and node memory states in CPU RAM. The training workflow proceeds as follows: ① **Sampling**: The CPU executes temporal neighborhood sampling to construct computation graphs. ② **Gathering**: Relevant features and states are extracted from host memory based on the sampled indices. ③ **Transfer**: The prepared mini-batch data is streamed to the GPU via PCIe. ④ **Compute**: The T-GNN model performs forward and backward passes on the GPU. ⑤ **State Update**: New node states are computed on the device. ⑥ **Synchronization**: Updated states are flushed back to CPU memory to maintain global consistency

exclusively on the GPU. However, this strategy becomes insufficient when dealing with large-scale dynamic graphs. For instance, the GDELT dataset [33], a real-world dynamic knowledge graph containing nearly a billion interactions, requires over 130GB of memory just to store the input data—far beyond the typical GPU memory capacity of 11 GB to 40 GB. To address this challenge, TGL [34] advocates a hybrid CPU-GPU data layout (illustrated in Fig. 2.3), which leverages the substantially larger CPU main memory (often exceeding 256GB) to store input data during training. Specifically, all input data are loaded in one batch from secondary storage into CPU memory immediately before training begins. Throughout training, the dynamic graph data, node state vectors, node features, and edge features remain in CPU memory. This approach aligns with the mainstream design of large-scale static GNN training systems [13, 35, 36]. Meanwhile, the GPU is dedicated to performing the model computations.

Preprocessing. Before training, the dynamic graph is preprocessed and divided into multiple batches that respect a temporal ordering constraint, as shown in Fig. 2.4. This constraint preserves the inherent temporal nature of the graph by ensur-

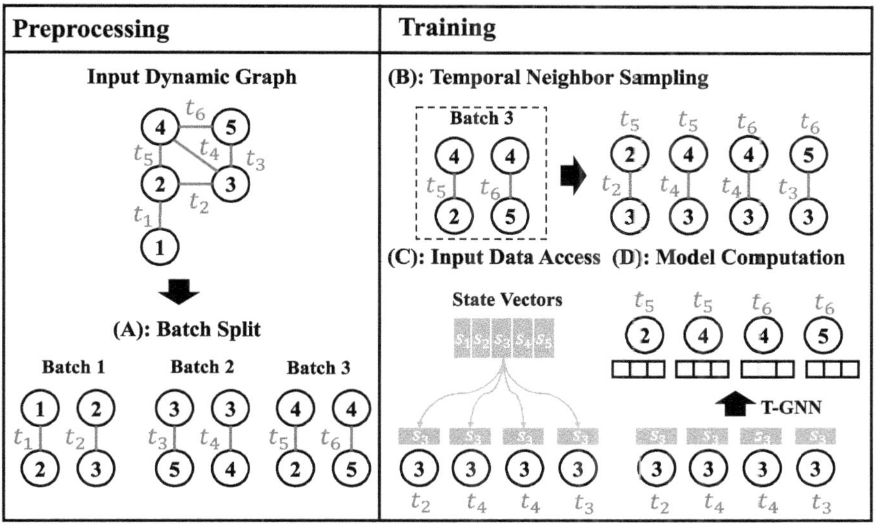

Fig. 2.4 Toy example of T-GNN training. The input dynamic graph contains 6 interactions with different timestamps. The batch size is 2, which means that 2 target interactions are contained in each batch. The number of sampled neighbor per node is 1 and the sampling depth is 1

ing that timestamp order is never violated. For example, swapping the interaction $(v_2, v_3, e_{23}(t_2), t_2)$ from Batch 1 with $(v_3, v_5, e_{35}(t_3), t_3)$ from Batch 2 is prohibited. Since T-GNN processes batches sequentially, preventing the use of future information guarantees fairness in prediction results.

Main Training Stages. Each training batch involves three main steps: temporal neighbor sampling, data retrieval, and model computation. For each target node, sampling is performed as in Eq. 2.10, but unlike static graphs, only neighbors with earlier timestamps are considered to avoid future information leakage. To improve efficiency, TGL [34] proposes a parallel sampler for simultaneous sampling across nodes. Based on the sampled subgraph, the corresponding node states, features, and edge features are retrieved and fed into the T-GNN model. To accelerate computation, Orca [37] reduces aggregation frequency via historical embeddings, while Zebra [38] modifies the aggregation rule to focus only on the most influential temporal neighbors [31].

2.4 Graph Neural Network Training Hardware Scenarios

2.4.1 Training on a Single GPU

When training GNNs on a single GPU, full-batch training (Fig. 2.5) is generally feasible only for small- to medium-scale graphs. Otherwise, the exponential neighbor growth causes memory and runtime bottlenecks. Hence, mini-batch training (Fig. 2.6)

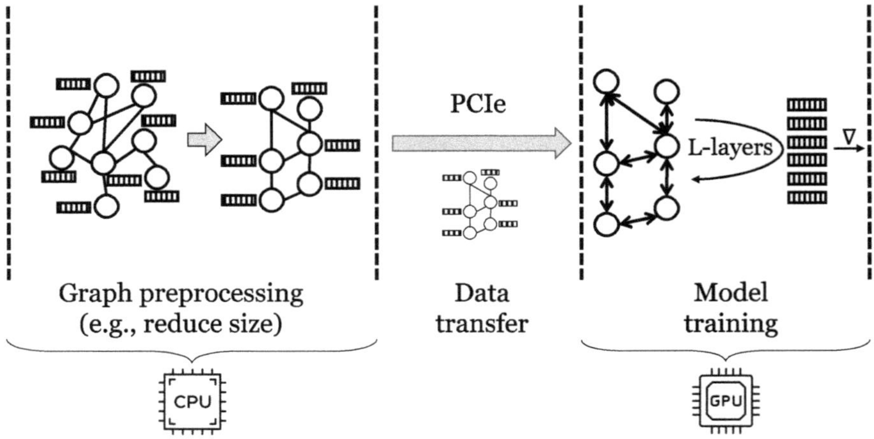

Fig. 2.5 Full graph training on a single GPU

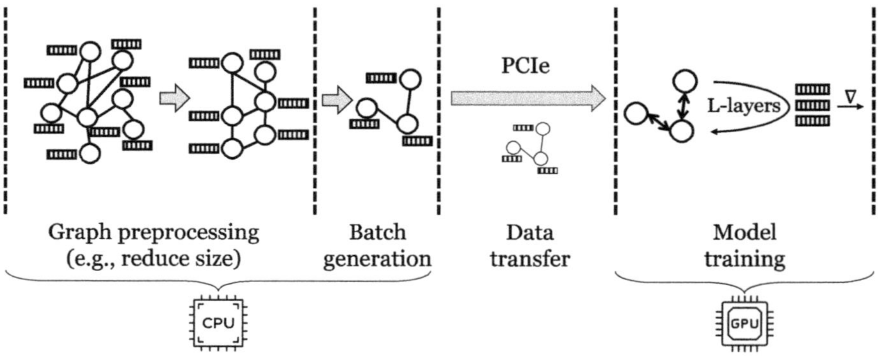

Fig. 2.6 Mini-batch training on a single GPU

with neighborhood sampling is commonly used to reduce computation and memory overhead. The basic workflow involves recursively sampling multi-hop neighbors from a set of target nodes to construct computation subtrees. Message passing and gradient updates are then performed on these subgraphs, repeating the process for the next batch.

Memory constraints are a major concern in single-GPU training. A common solution is to preload the training graph into CPU memory and use an asynchronous pipeline to load subgraph structures and node features onto the GPU for forward and backward passes. Libraries such as PyTorch Geometric and DGL support these sampling-based mini-batch training methods. Further optimizations include layer-wise caching or feature chunking to reduce redundant reads and data transfer. Overall, single-GPU training leverages various sampling and subgraph techniques to miti-

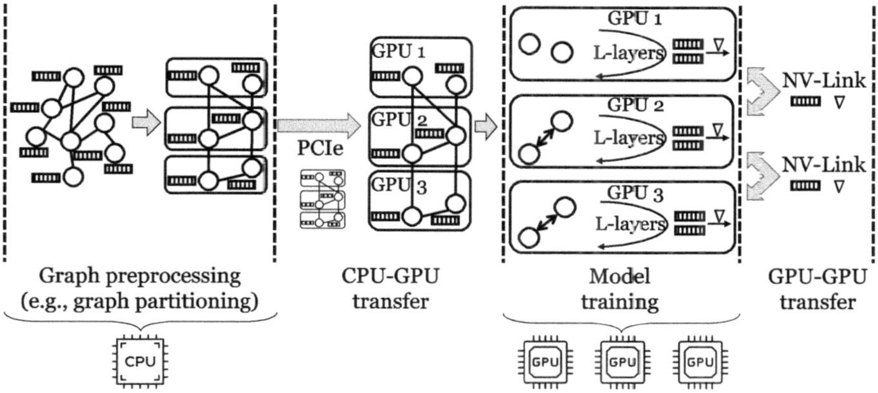

Fig. 2.7 Full graph training on multiple GPUs

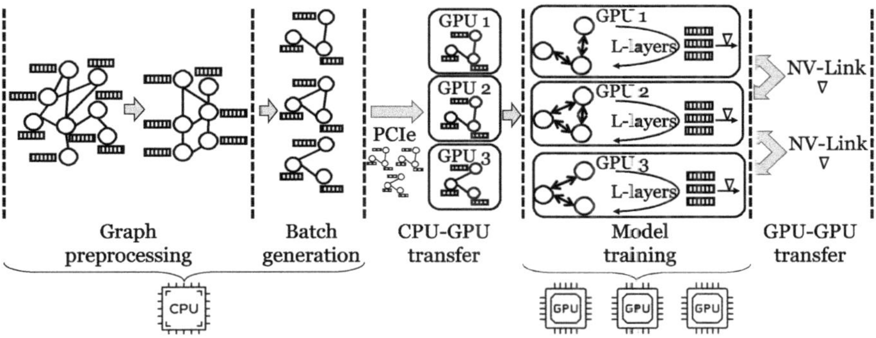

Fig. 2.8 Mini-batch training on multiple GPUs

gate computational and memory demands, albeit introducing a trade-off between sampling-induced variance and model expressiveness.

2.4.2 *Training in Multi-GPU Environments*

In multi-GPU environments, parallelism can further scale up training capacity. Two main strategies are used: model parallelism that distributes different layers or parameter subsets across GPUs and data parallelism that replicates the model across GPUs and distributes input data. Given the complex computation graph of GNNs, data parallelism is more common. The main challenges include inter-GPU communication and workload balancing. Communication can be handled via high-speed interconnects (e.g., NVLink, PCIe P2P) or NCCL-based AllReduce synchronization, or using a parameter server to coordinate gradient updates. Workload balance depends on effective graph partitioning to distribute subgraphs while minimizing inter-GPU neighbor access [39–41]. Figures 2.7 and 2.8 show procedures of data parallel GNN training on multiple GPUs.

In summary, multi-GPU training extends GNN scalability through parallelism. Key considerations include efficient partitioning of graphs/models and minimizing inter-GPU communication. Compared to single-GPU setups, additional GPUs boost throughput but introduce new challenges such as communication overhead and redundant data storage (e.g., cross-device caching). Sampling strategies in multi-GPU setups are also more flexible since each GPU can independently perform neighbor sampling or process different subgraphs to improve parallel efficiency.

2.4.3 Distributed Training on GPU Clusters

Training GNNs on large-scale GPU clusters introduces additional challenges involving cross-node communication, synchronization, and distributed algorithm design. Distributed frameworks generally adopt either synchronous training (e.g., AllReduce-based parameter synchronization per batch) or asynchronous training (e.g., parameter server-based gradient updates). Synchronous training maintains strict model consistency but suffers from communication overhead, while asynchronous training is more latency-tolerant but requires conflict resolution and convergence handling.

Cross-node communication is a major bottleneck in distributed GNNs. Unlike dense tensors in traditional DNNs, GNNs require transmission of sparse graph structures and features across nodes. For instance, when a node's neighbors span multiple machines, their feature vectors must be fetched over the network, leading to numerous sparse requests. Some systems alleviate this by preloading batches and fusing requests over high-speed intra-server links. Under the parameter server (PS) architecture, servers store node embeddings or global parameters, and workers push/pull gradients during training. In contrast, AllReduce architectures use MPI/NCCL libraries for periodic gradient synchronization (e.g., Horovod). For example, DistDGL [42] initially adopted a CPU-based PS design, supporting distributed storage and updates of embeddings. Systems like GNNLab [43] prefer AllReduce for consistent multi-GPU training.

Graph partitioning and redistribution are critical for efficient distributed GNN training. Typical approaches divide the graph into subgraphs assigned to different nodes, each training on its own subgraph while periodically exchanging border information. Classical partitioning methods include vertex- and edge-cut algorithms (e.g., METIS, ParMETIS) as well as streaming partitioning algorithms. Graph partitioning reduces communication and balances load but introduces challenges in synchronizing boundary nodes. Some systems cache features of cross-partition neighbors to accelerate access and apply consistency protocols during graph updates. For dynamic graphs, recent systems adaptively update partitions or caching policies during training, though most current systems assume static graph structures.

Distributed sampling is another direction to improve efficiency. A common approach is for multiple nodes to independently sample neighborhoods or subgraphs in parallel, then proceed with local computation.

References

1. Thomas N. Kipf, Max Welling. 2017. Semi-Supervised Classification with Graph Convolutional Networks. In *ICLR*.
2. Petar Velickovic, Guillem Cucurull, Arantxa Casanova, Adriana Romero, Pietro Liò, Yoshua Bengio. 2018. Graph Attention Networks. In *ICLR*.
3. Thomas N. Kipf and Max Welling. 2017. Semi-Supervised Classification with Graph Convolutional Networks. In *ICLR*. https://openreview.net/forum?id=SJU4ayYgl
4. Sergi Abadal, Akshay Jain, Robert Guirado, Jorge López-Alonso, and Eduard Alarcón. 2022. Computing Graph Neural Networks: A Survey from Algorithms to Accelerators. *ACM* Comput. Surv. 54, 9 (2022), 191:1–191:38. https://doi.org/10.1145/3477141
5. Zonghan Wu, Shirui Pan, Fengwen Chen, Guodong Long, Chengqi Zhang, and Philip S. Yu. 2021. A Comprehensive Survey on Graph Neural Networks. *IEEE* Trans. Neural Networks Learn. Syst. 32, 1 (2021), 4–24.
6. Qiqi Zhou, Yanyan Shen, Lei Chen. 2023. Narrow the Input Mismatch in Deep Graph Neural Network Distillation. In *SIGKDD*. 3581–3592. https://doi.org/10.1145/3580305.3599442
7. Qiqi Zhou, Yanyan Shen, Lei Chen. 2025. Faster Convergence in Mini-batch Graph Neural Networks Training with Pseudo Full Neighborhood Compensation. *PVLDB*. 18(11): 4309–4322.
8. William L. Hamilton, Zhitao Ying, Jure Leskovec. 2017. Inductive Representation Learning on Large Graphs. In *NIPS*. 1024–1034.
9. Rex Ying, Ruining He, Kaifeng Chen, Pong Eksombatchai, William L. Hamilton, Jure Leskovec. 2018. Graph convolutional neural networks for web-scale recommender systems. In *KDD*. 257–266.
10. Matthias Fey, Jan Eric Lenssen. 2019. Fast graph representation learning with PyTorch Geometric. In *ICLR Workshop*.
11. Difan Zou, Ziniu Hu, Yewen Wang, Song Jiang, Yizhou Sun, Quanquan Gu. 2019. Layer-dependent importance sampling for training deep and large graph convolutional networks. *NeurIPS*. 11247–11256.
12. Hanqing Zeng, Hongkuan Zhou, Ajitesh Srivastava, Rajgopal Kannan, Viktor K. Prasanna. 2020. GraphSAINT: Graph sampling based inductive learning method. In *ICLR*.
13. Jianbang Yang, Dahai Tang, Xiaoniu Song, Lei Wang, Qiang Yin, Rong Chen, Wenyuan Yu, and Jingren Zhou. 2022. GNNLab: a factored system for sample-based GNN training over GPUs. In *EuroSys*. 417–434.
14. Wei-Lin Chiang, Xuanqing Liu, Si Si, Yang Li, Samy Bengio, Cho-Jui Hsieh. 2019. Cluster-GCN: An efficient algorithm for training deep and large graph convolutional networks. In *KDD*. 257–266.
15. Prithviraj Sen, Galileo Namata, Mustafa Bilgic, Lise Getoor, Brian Gallagher, Tina Eliassi-Rad. 2008. Collective Classification in Network Data. *AI Mag*. 29(3), 93–106.
16. Jure Leskovec, Julian McAuley. 2012. Social circles: Facebook. *Stanford Network Analysis Project (SNAP)*. https://snap.stanford.edu/data/ego-Facebook.html
17. Damian Szklarczyk, Annika L. Gable, Katerina C. Nastou, David Lyon, Rebecca Kirsch, Sampo Pyysalo, Nadezhda T. Doncheva, Marc Legeay, Tao Fang, Peer Bork, Lars Juhl Jensen, Christian von Mering. 2021. The STRING database in 2021: customizable protein-protein networks, and functional characterization of user-uploaded gene/measurement sets. *Nucleic Acids Res*. **49**(D1), D605–D612 (2021). https://string-db.org/

18. Reza Zafarani, Huan Liu. Social computing data repository: BlogCatalog. 2009. *ASU Social Computing Data Repository*. http://socialcomputing.asu.edu/datasets/BlogCatalog3
19. Pietro Panzarasa, Tore Opsahl, Kathleen M. Carley. 2009. UCI Online Social Network Dataset. *UC Irvine Mach. Learn. Repos*. https://archive.ics.uci.edu/ml/datasets/Online+Social+Network
20. Asim Kumar Debnath, Rosa L Lopez de Compadre, Gargi Debnath, Alan J Shusterman, Corwin Hansch. 1991. Structure-activity relationship of mutagenic aromatic and heteroaromatic nitro compounds. Correlation with molecular orbital energies and hydrophobicity. *J. Med. Chem.* **34**(2), 786–797.
21. Karsten M. Borgwardt, Cheng Soon Ong, Stefan Schönauer, S. V. N. Vishwanathan, Alexander J. Smola, Hans-Peter Kriegel. 2005. Protein function prediction via graph kernels. *ISMB*. 47–56.
22. Nikil Wale, Ian A. Watson, George Karypis. 2008. Comparison of descriptor spaces for chemical compound retrieval and classification. *Knowl. Inf. Syst.* **14**(3), 347–375.
23. Zhilin Yang, William W. Cohen, Ruslan Salakhutdinov. 2016. Revisiting semi-supervised learning with graph embeddings. In *ICML*. 40–48.
24. Matthias Fey, Jan Eric Lenssen. 2019. Fast graph representation learning with PyTorch Geometric. In *ICLR Workshop*.
25. Minjie Wang, Da Zheng, Zihao Ye, Quan Gan, Mufei Li, Xiang Song, Jinjing Zhou, Chao Ma, Lingfan Yu, Yu Gai, et al. 2019. Deep graph library: A graph-centric, highly-performant package for graph neural networks. *arXiv preprint* arXiv:1909.01315 (2019).
26. Da Xu, Chuanwei Ruan, Evren Körpeoglu, Sushant Kumar, Kannan Achan. 2020. Inductive representation learning on temporal graphs. In *ICLR*.
27. Emanuele Rossi, Ben Chamberlain, Fabrizio Frasca, Davide Eynard, Federico Monti, Michael M. Bronstein. 2020. Temporal Graph Networks for Deep Learning on Dynamic Graphs. In *ICML 2020 Workshop on Graph Representation Learning*.
28. Yanbang Wang, Yen-Yu Chang, Yunyu Liu, Jure Leskovec, Pan Li. 2021. Inductive Representation Learning in Temporal Networks via Causal Anonymous Walks. In *ICLR*.
29. Xuhong Wang, Ding Lyu, Mengjian Li, Yang Xia, Qi Yang, Xinwen Wang, Xinguang Wang, Ping Cui, Yupu Yang, Bowen Sun, Zhenyu Guo. 2021. Apan: Asynchronous propagation attention network for real-time temporal graph embedding. In *Proceedings of the 2021 International Conference on Management of Data*. pp. 2628–2638.
30. Srijan Kumar, Xikun Zhang, Jure Leskovec. 2019. Predicting dynamic embedding trajectory in temporal interaction networks. In *SIGKDD*. pp. 1269–1278.
31. Emanuele Rossi, Ben Chamberlain, Fabrizio Frasca, Davide Eynard, Federico Monti, Michael M. Bronstein. 2020. Temporal Graph Networks for Deep Learning on Dynamic Graphs. In *ICML 2020 Workshop on Graph Representation Learning*.
32. Yiming Li, Yanyan Shen, Lei Chen, and Mingxuan Yuan. 2023b. Zebra: When Temporal Graph Neural Networks Meet Temporal Personalized PageRank. *VLDB* 16, 6 (2023), 1332–1345.
33. Leetaru Kalev, Philip A. Schrodt. 2013. Gdelt: Global data on events, location, and tone, 1979–2012. In *ISA Annual Convention*. vol. 2, no. 4, 1–49.
34. Hongkuan Zhou, Da Zheng, Israt Nisa, Vasileios Ioannidis, Xiang Song, George Karypis. 2022. TGL: A General Framework for Temporal GNN Training on Billion-Scale Graphs. *PVLDB*.
35. Xin Zhang, Yanyan Shen, Yingxia Shao, and Lei Chen. 2023. DUCATI: A Dual-Cache Training System for Graph Neural Networks on Giant Graphs with the GPU. *Proc. ACM Manag. Data* 1, 2 (2023), 166:1–166:24. https://doi.org/10.1145/3589311
36. Zhiqi Lin, Cheng Li, Youshan Miao, Yunxin Liu, and Yinlong Xu. 2020. Pagraph: Scaling gnn training on large graphs via computation-aware caching. In *SoCC*. 401–415.
37. Yiming Li, Yanyan Shen, Lei Chen, Mingxuan Yuan. Orca: Scalable Temporal Graph Neural Network Training with Theoretical Guarantees. *Proc. ACM Manag. Data*. 1(1): 52:1–52:27.
38. Yiming Li, Yanyan Shen, Lei Chen, Mingxuan Yuan. 2023. Zebra: When Temporal Graph Neural Networks Meet Temporal Personalized PageRank. *PVLDB*. 16(6): 1332–1345.

39. Zhiqi Lin, Cheng Li, Youshan Miao, Yunxin Liu, Yinlong Xu. 2020. Pagraph: Scaling GNN Training on Large Graphs via Computation-Aware Caching. In *SoCC*. 401–415.
40. Rong Zhu, Kun Zhao, Hongxia Yang, Wei Lin, Chang Zhou, Baole Ai, Yong Li, Jingren Zhou. 2019. AliGraph: A Comprehensive Graph Neural Network Platform. *PVLDB*. 12(12): 2094–2105.
41. Lingxiao Ma, Zhi Yang, Youshan Miao, Jilong Xue, Ming Wu, Lidong Zhou, Yafei Dai. 2019. NeuGraph: Parallel Deep Neural Network Computation on Large Graphs. In *ATC*. 443–458.
42. Da Zheng, Xiang Song, Chengru Yang, Dominique LaSalle, George Karypis. 2022. Distributed Hybrid CPU and GPU Training for Graph Neural Networks on Billion-Scale Heterogeneous Graphs. In *SIGKDD*. 4582–4591.
43. Jianbang Yang, Dahai Tang, Xiaoniu Song, Lei Wang, Qiang Yin, Rong Chen, Wenyuan Yu, Jingren Zhou. 2022. GNNLab: A Factored System for Sample-Based GNN Training over GPUs. In *EuroSys*. 417–434.

Chapter 3
Reducing Data Transfer for Scalable Graph Neural Network Training

Xin Zhang

Abstract Graph Neural Networks are popular these days due to their strong capability in modeling and predicting complex graph-structured data. However, real-world graphs are usually giant in nature, containing millions of nodes and billions of edges. To enhance the training scalability, many mini-batch GNN training methods have been proposed recently. However, we observe two significant gaps that hinder the efficient and scalable training of GNNs. In this chapter, we first provide a structured review of these existing works and categorize them based on their position in the GNN training stack, namely the graph sampling algorithm and the GNN training system. Then, we analyze the shortcomings of existing work and propose our solutions, namely a scalable graph sampling algorithm called feature-oriented sampling and an efficient GNN training system called DUCATI to address them.

3.1 Background

GNNs are featured by the recursive information aggregation and neural transformation of node features within the graph topology. Therefore, unlike traditional deep learning, where training samples are independent, samples in a graph dataset are dependent on other samples during training. Traditional GNNs [1] even require that the whole dataset must be loaded on to the GPU during training.

Unfortunately, many real-world graphs are giant by nature, making training GNNs challenging. For example, the Ogbn-Papers100M dataset [2], a citation graph of papers published on the arXiv, contains over 111 million nodes and 1.6 billion edges. The adjacency data and the node features take around 80GB of storage, far exceeding the dozen GB or so memory capacity of consumer-level GPUs. As a result, mini-batch training has become the de facto way to train GNNs on giant graphs. Given a large graph and corresponding node features, we sample the input graph to get a subgraph and then select the corresponding node features for the subgraph. From the perspective of the underlying data structure, we conduct sampling on the adjacency

X. Zhang (✉)
The Hong Kong University of Science and Technology, Hong Kong SAR, China
e-mail: xzhanggb@connect.ust.hk

Y. Shen and L. Chen (eds.), *Graph Neural Network Training*, Machine Learning: Foundations, Methodologies, and Applications,
https://doi.org/10.1007/978-981-95-5795-0_3

matrix and perform selection on the node features tensor. The subgraph, together with its node features, forms a mini-batch that will be transferred to the GPU for later training.

At the core of efficient and scalable GNN training are two important components, namely the graph sampling algorithm and the system design for GNN training. The former determines the actual batch size and content and balances the trade-off between accuracy and scalability in iterative GNN training. The latter determines how the graph sampling, feature gathering, and model-training stages are executed on CPU, PCIe, and GPU, while maximizing the hardware utilization. In recent years, many efforts have been made in the two areas to accelerate the GNN training.

First, many graph sampling algorithms are proposed in recent works [3–8]. Depending on the sampling operand and output, mainstream sampling algorithms can be categorized into three types, namely the node-wise sampling, the layer-wise sampling, and the subgraph-wise sampling. Given the target nodes to be predicted, node-wise sampling algorithms directly sample the neighbors of the target nodes layer by layer. On the other hand, the layer-wise sampling algorithms are conducted on each GNN layer by excluding a part of the participating nodes based on certain criteria. The subgraph-wise sampling algorithms extract subgraphs from the input dataset using methods like random walking or partitioning. These works significantly reduce the size of the related nodes, and we only need to maintain the sampled neighbors instead of the whole graph on GPU during training, which drastically improves scalability. Various sampling algorithms have different pros and cons, and we will introduce and compare them in detail in the following sections.

Second, many system works [9–15] have been published to enhance the efficiency and improve the hardware utilization of GNN training. While the aforementioned sampling works indeed reduce the data volume to some extent, the data transferring between CPU and GPU or between GPUs is still significant and hinders the GNN training process. To address this problem, researchers propose different ways to either (1) reduce the data transfer volume or (2) improve the bandwidth of data transfer. For example, cache-based methods and staleness-based methods directly reduce the data transfer volume using exact or approximate caching. The UVA-based methods and GPUDirect-based methods improve the bandwidth utilization by letting GPU send fine-grained data-access requests directly to main memory or SSD without being blocked by CPU. There are also many works that design sophisticated execution strategies, such as pipeline execution in PipeGCN [16] and factored execution in GNNLab [11]. These works are suitable under different settings and also have their own strengths and weaknesses.

3.2 Feature-Oriented Sampling

3.2.1 Drawbacks of Topology-Oriented Sampling

As discussed previously, most existing graph sampling strategies fall into three classes: node-wise, layer-wise, and subgraph-wise sampling. We collectively denote these approaches as Topology-Oriented Sampling (TOS) methods, and summarize their mini-batch construction process in Fig. 3.1. A graph dataset consists of two fundamental parts: the topology component (i.e., the input graph shown on the left of Fig. 3.1) and the feature component (i.e., vertex features illustrated in the middle of Fig. 3.1). In widely used GNN training libraries such as PyTorch-Geometric [5] and DGL [17], the feature component is stored as a large two-dimensional tensor, where the ith row corresponds to the feature vector of vertex i. Each mini-batch also contains these two components, albeit at a significantly smaller scale. During mini-batch construction, TOS methods first decide the topology component and then generate the associated feature component. Concretely, they begin by forming the topology of a mini-batch through sampling the L-hop neighbors (where L denotes the model depth) of the training nodes from the input graph. Different samplers adopt different strategies for this step. For instance, the SAGE sampler applies node-level sampling by limiting the number of neighbors per node, while the SAINT sampler conducts subgraph-level sampling by inducing a subgraph from selected nodes. Once the topology component is obtained, TOS methods gather the indices of all vertices contained in the sampled subgraph. They then construct the feature component by extracting the corresponding rows from the large 2-D feature tensor using these indices and copying them into a contiguous buffer. Finally, the assembled mini-batch is moved to GPU memory to execute one training iteration.

However, TOS methods exhibit three notable limitations. (1) The vertex indices in a sampled subgraph are often highly irregular, leading to frequent random accesses to graph data (both topology and features) during mini-batch preparation. This behavior imposes heavy I/O overhead on the CPU and substantially slows down training. As an

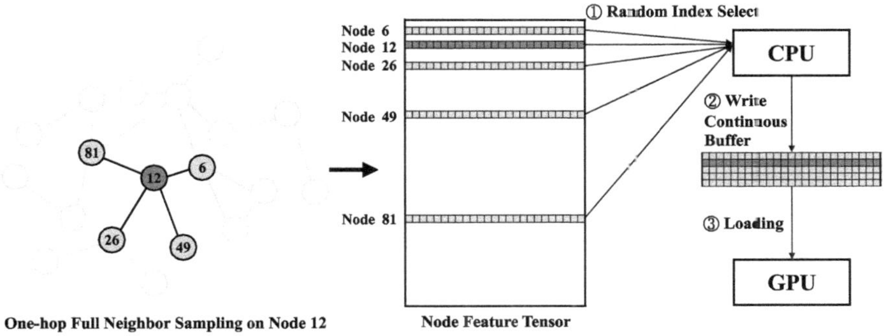

Fig. 3.1 Topology-oriented sampling: first sample a subgraph and then fetch node features

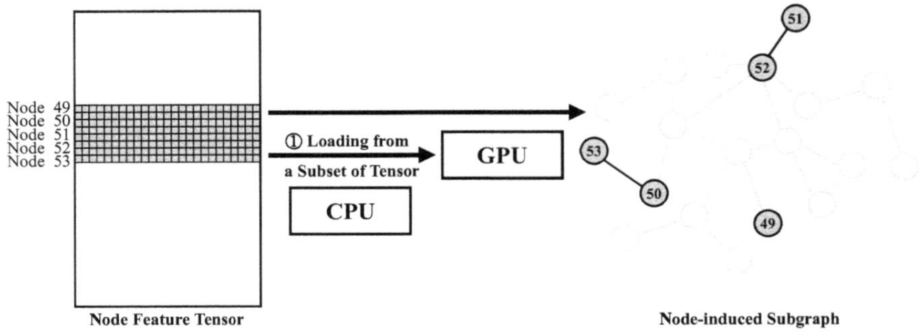

Fig. 3.2 Feature-oriented sampling: first sample a block of node features and then induce a subgraph

example, when running GraphSAINT on the Amazon dataset, the *aten::index_select* operator becomes the dominant performance bottleneck, as reported in Table 3.1. (2) As model depth increases, the size of mini-batches produced by the SAGE sampler or the full neighbor sampler grows exponentially, which negatively impacts scalability and training efficiency. For instance, on the Amazon dataset with 1,255,968 training nodes, increasing L from 2 to 3 raises the average number of sampled nodes from 54,385 to 132,213 for the SAGE sampler, and from 651,540 to 803,259 for the full neighbor sampler. (3) From a sampling standpoint, TOS methods restrict sampled nodes to the L-hop neighborhood of the training set, which typically covers only a subset of the entire graph. Consequently, under the transductive learning setting, these methods are unable to exploit features from all input nodes. This limitation introduces non-trivial information loss and can degrade model performance. For example, on the Ogbn-Papers100M dataset, the coverage ratios (number of nodes in the L-hop neighborhood of training set/total number of nodes in the input graph) are merely 0.039, 0.107, 0.230, and 0.362 when L is set to 1, 2, 3, and 4, respectively.

The drawbacks discussed above naturally motivate the following question: can we (1) remove random accesses to graph data, (2) prevent the mini-batch size from growing exponentially, and (3) utilize all available input features during sampling when training GNNs on large-scale graphs? In this section, we introduce the FOS sampler as a positive response to these challenges. The FOS sampler is a Feature-Oriented Sampling approach designed for efficient GNN training. An overview of the FOS sampler is shown in Fig. 3.2. In contrast to topology-oriented methods, our approach first fixes the feature component of a mini-batch and subsequently derives its topology component. Specifically, we begin by choosing a contiguous sub-tensor from the input node feature tensor to serve as the feature component of the mini-batch, as illustrated on the left side of Fig. 3.2. The indices corresponding to these selected node features are then obtained, which are inherently consecutive. Based on these indices, we construct the topology component by extracting the node-induced subgraph from the original input graph, as depicted on the right side of Fig. 3.2. Finally, the completed mini-batch is transferred to GPU memory for subsequent training (Table 3.2).

Table 3.1 Top 10 most time-consuming operators when training GraphSAINT on the Amazon dataset

Operator name	Duration (ms)	Self time (ms)	Avg. Duration (ms)	Occurrences
aten::index_select	213.660	213.044	3.142	68
aten::to	102.923	2.403	0.165	524
aten::copy_	99.546	1.007	1.555	64
cudaMemcpyAsync	91.313	91.313	0.878	104
cudaStreamSynchronize	48.046	48.046	0.632	76
Optimizer.step#Adam.step	32.320	9.744	8.080	4
cudaLaunchKernel	31.234	31.234	0.011	2948
aten::is_nonzero	26.947	0.044	2.246	12
aten::item	26.903	0.056	2.242	12
aten::_local_scalar_dense	26.847	0.092	2.237	12

We use the RandomWalkSampler of GraphSAINT with num_roots = 3000 and length = 3. The first 4 iterations are warm-up stage and we measure the accumulated training time of the following 4 iterations with PyTorch Profiler [18]

Because the feature component of each mini-batch is stored contiguously, random access to node features is completely avoided. Moreover, since node identifiers are consecutive, random access to topology data is also largely mitigated, as the associated structural information is typically stored in a contiguous manner. As a result, the FOS sampler significantly reduces memory I/O overhead and effectively addresses the first limitation of TOS methods. Furthermore, the FOS sampler decouples the sampling process from the model depth, ensuring that the mini-batch size remains independent of the number of GNN layers. As demonstrated experimentally in Sect. 3.3, the resulting batch sizes are consistently smaller and grow much more slowly than those produced by topology-oriented approaches. Finally, as detailed in Sect. 3.2.2, the FOS sampler uniformly exploits all input node features, with every node in the input graph considered as a potential sampling candidate.

3.2.2 *GNNs with Feature-Oriented Sampling*

The FOS sampler is compatible with a wide range of GNN architectures under both inductive and transductive learning scenarios. The fundamental principle behind feature-oriented sampling is to build message-passing subgraphs whose nodes occupy continuous index intervals. As discussed in Sect. 3.1, graph data associated with consecutively indexed nodes reside in a compact memory layout, enabling efficient transfer to GPU memory during training. When incorporating the FOS sampler into standard GNN models, the primary modification lies in replacing the conventional mini-batch construction strategy. In the remainder of this part, we demonstrate

Algorithm 1 Training algorithm for the vanilla FOS-SAINT.

Input: Training graph $\mathcal{G}$ $(\mathcal{V}, \mathcal{E}, \boldsymbol{X})$; Initial model weights
Output: Trained model weights

Preprocessing: Randomly permuted G; Compute normalization coefficients.
for e **do**ach mini-batch
    $\mathcal{G}_s$ $(\mathcal{V}_s, \mathcal{E}_s, \boldsymbol{X}_s)$ $\leftarrow$ Randomly choose a start index i; use nodes with indices in range$(i, i+b)$ to induce a subgraph of $\mathcal{G}$ as the topology component; use node features with the same indices as the feature component.
    Load model weights from the last step and perform the forward computation and backward propagation on $\mathcal{G}_s$ $(\mathcal{V}_s, \mathcal{E}_s, \boldsymbol{X}_s)$ with normalization.
    Update the model weights.
end for

how the FOS sampler is integrated into representative GNN models. In particular, we describe the vanilla FOS-SAINT in Sect. 3.2.2 as a representative case of inductive GNN training, and the vanilla FOS-GCN in Sect. 3.2.2 as an example under the transductive setting. To mitigate the limited receptive field issue inherent in the vanilla FOS sampler, we further present an enhancement in Sect. 3.2.2, which is applicable to both inductive and transductive tasks. In addition, Sect. 3.2.2 outlines practical guidelines for selecting key hyperparameters, informed by theoretical analysis and system considerations. Finally, in Sect. 3.2.2, we provide a theoretical investigation of the FOS sampler and a comparative analysis of the properties of subgraphs sampled by the FOS sampler versus those produced by alternative sampling methods.

Vanilla FOS-SAINT for Inductive Learning

In the inductive learning scenario, vertices belonging to the validation and test sets are not involved during training. For convenience, we denote by $\mathcal{G}$ the subgraph induced by all training vertices from the original input graph. We are thus given an attributed graph $\mathcal{G}$ $(\mathcal{V}, \mathcal{E}, \boldsymbol{X})$, where $N = |\mathcal{V}|$ denotes the number of training vertices and $|\mathcal{E}|$ denotes the number of edges. The matrix $\boldsymbol{X} \in \mathbb{R}^{|\mathcal{V}| \times |F|}$ stores the vertex features, with the ith row representing the F-dimensional feature vector of vertex i. Prior to training, the attributed graph $\mathcal{G}$ is loaded and stored in main memory.

The overall training workflow is summarized in Algorithm 1. Before the training procedure begins, we first apply a random permutation to $\mathcal{G}$ in order to mitigate any potential bias arising from the dataset construction. The feature matrix $\boldsymbol{X}$ is permuted accordingly to remain consistent with the reordered graph. During each training iteration, FOS-SAINT randomly accesses a block of node features that forms a contiguous segment of $\boldsymbol{X}$. Based on the selected vertices, a node-induced subgraph is then constructed. Subsequently, the GCN model is trained on this subgraph with the appropriate normalization applied. At each iteration, the GCN is optimized using only a randomly sampled subgraph. Nevertheless, as training progresses, information from different parts of the graph is gradually incorporated, enabling the model to converge to a well-trained GCN over the entire graph. The sampling strategy described above results in non-uniform probabilities for edges to be included in individual mini-batches, which may introduce bias into the GCN training process. To counteract this effect, we adopt a normalization strategy inspired by GraphSAINT.

Due to the presence of non-linear activation functions between GCN layers, directly analyzing the induced bias is challenging. Therefore, we focus on the aggregation operation between two consecutive layers prior to the activation function. This analytical approach has also been employed in prior studies [3, 19]. Let $\boldsymbol{x}_v$ denote the feature vector of vertex v, $\boldsymbol{A}$ denote the full adjacency matrix, and $\widetilde{\boldsymbol{A}} = \boldsymbol{D}^{-1}\boldsymbol{A}$ denote the normalized adjacency matrix, where $\boldsymbol{D}$ is the diagonal degree matrix.

Our objective is to guarantee that, for every vertex v, the aggregated feature representation ϕ_v computed on a sampled subgraph serves as an unbiased estimator of the aggregation over the entire graph, which we denote by φ_v. Let $f^{(\ell)}$ represent the dimensionality of the input activation at layer ℓ. The activation vector of vertex v at layer ℓ is $\boldsymbol{x}_v^{(\ell)} \in \mathbb{R}^{f^{(\ell)}}$, and the corresponding weight matrix is $\boldsymbol{W}^{(\ell)} \in \mathbb{R}^{f^{(\ell)} \times f^{(\ell+1)}}$.

For a standard GCN operating on the full graph, the aggregated features $\varphi_v^{\ell+1}$ of vertex v at layer $(\ell + 1)$ are computed as:

$$\varphi_v^{(\ell+1)} = \sum_{u \in \mathcal{V}} \widetilde{\boldsymbol{A}}_{v,u} \left(\boldsymbol{W}^{(\ell)}\right)^{\mathsf{T}} \boldsymbol{x}_u^{(\ell)}. \tag{3.1}$$

To approximate φ_v using a sampled subgraph, we introduce an aggregation normalization coefficient $\alpha_{u,v}$. With this coefficient, the aggregated representation $\phi_v^{\ell+1}$ is computed as follows:

$$\phi_v^{(\ell+1)} = \sum_{u \in \mathcal{V}} \frac{\widetilde{\boldsymbol{A}}_{v,u}}{\alpha_{u,v}} \left(\boldsymbol{W}^{(\ell)}\right)^{\mathsf{T}} \boldsymbol{x}_u^{(\ell)} \mathbb{1}_{u|v}. \tag{3.2}$$

Here, $\mathbb{1}_{u|v} \in \{0, 1\}$ denotes an indicator variable specifying whether the edge (u, v) is present in the sampled subgraph. Concretely, $\mathbb{1}_{u|v} = 0$ if $v \in \mathcal{V}_s \wedge (u, v) \notin \mathcal{E}_s$; $\mathbb{1}_{u|v} = 1$ if $(u, v) \in \mathcal{E}_s$; and $\mathbb{1}_{u|v}$ is undefined when $v \notin \mathcal{V}_s$.

In order for $\phi_v^{(\ell+1)}$ to serve as an unbiased estimator of $\varphi_v^{(\ell+1)}$, it is required that $\mathbb{E}\left(\phi_v^{(\ell+1)}\right) = \varphi_v^{(\ell+1)}$. Under the FOS-SAINT scheme, the input graph is first randomly permuted at the node level, which allows the vertices sampled at each iteration to be regarded as uniformly drawn from the entire vertex set. This property enables us to directly derive $\alpha_{u,v}$ as $\alpha_{u,v} = \mathbb{E}\left(\mathbb{1}_{u|v}\right) = p(u|v) = 1/\text{in\_degrees}(v)$.

We require that the loss computed on each mini-batch be an unbiased estimator of the loss over the full graph. To this end, we introduce a loss normalization coefficient λ_v for each vertex v when computing the training loss. Let L_v denote the loss associated with vertex v at the output layer, and let b represent the batch size of the sampled subgraph. For a normal GCN on the whole graph, the loss L_{full} is calculated by:

$$L_{\text{full}} = \frac{1}{N} \sum_{v \in \mathcal{G}} L_v.$$

Algorithm 2 Training algorithm for the vanilla FOS-GCN.

Input: Whole graph $\mathcal{G}\ (\mathcal{V}, \mathcal{E}, \boldsymbol{X})$; Initial GCN weights
Output: Trained GCN weights
1: Preprocessing: Randomly permuted G;
2: **for** each minibatch **do**
3: $\mathcal{G}_s\ (\mathcal{V}_s, \mathcal{E}_s, \boldsymbol{X}_s) \leftarrow$ Randomly choose a start index i; use nodes with indices in range$(i, i+b)$ to induce a subgraph of $\mathcal{G}$ as the topology component; use node features with the same indices as the feature component.
4: Load model weights from the last step and perform the forward computation on $\mathcal{G}_s\ (\mathcal{V}_s, \mathcal{E}_s, \boldsymbol{X}_s)$.
5: Identify the training nodes in the current subgraph, calculate loss based on them, and perform backward propagation.
6: Update the model weights.
7: **end for**

The loss on the sampled subgraph L_{batch} after normalization is given by:

$$L_{\text{batch}} = \frac{1}{b} \sum_{v \in \mathcal{G}_s} \frac{L_v}{\lambda_v}.$$

Let $\mathbb{E}\left(L_{\text{batch}}\right) = L_{\text{full}}$, we have $\lambda_v = p_v \cdot N = b$. The two normalization coefficients required by FOS-SAINT can be straightforwardly determined once the graph dataset and the relevant training hyperparameters are specified. For FOS-SAGE, the mini-batch construction procedure is identical to that of FOS-SAINT; however, the underlying model is substituted with GraphSAGE, and no normalization is applied during training.

Vanilla FOS-GCN for Transductive Learning

In the transductive learning setting, feature information from validation and test nodes can be exploited during training. Consequently, instead of performing sampling on the subgraph induced by training nodes only, we conduct sampling over the entire graph to construct mini-batches. For simplicity, we continue to denote the input graph as $\mathcal{G}$. Since label information is available only for training nodes within each mini-batch, it is necessary to identify these training nodes after the forward pass and compute the loss and gradients solely based on them. Before training, the whole graph is randomly permuted so that training nodes are uniformly distributed among all vertices. This strategy ensures that each mini-batch contains training nodes, provided that the training set ratio and batch size are sufficiently large. The overall training procedure is summarized in Algorithm 2.

FOSSampler with Decomposition Enhancement

Although the vanilla FOS-GNN is capable of delivering good performance, as will be demonstrated later, it still encounters the issue of a limited receptive field. Specifically, since each mini-batch is formed from nodes within consecutive index ranges, the potential neighbors of the ith vertex across all mini-batches are confined to a narrow interval, namely $[i - b + 1, i + b)$. Nodes outside this interval will never co-occur with the ith vertex in the same mini-batch, and thus cannot directly exchange information with it. Formally, if (u, v) denotes an edge between node u and node v, then all edges included in a sampled subgraph must satisfy $|u - v| < b$, as both endpoints are drawn from a consecutive range of size b. Any edge with $|u - v| \geq b$ will never appear in a sampled subgraph. This restriction limits information propagation across the graph and negatively affects model performance.

To overcome this limitation, we introduce a decomposition enhancement for the vanilla FOS-GNN, which effectively enlarges the receptive field of each node to encompass all other nodes in the graph. Concretely, we decompose the original node interval $[i, i + b)$ into d independently and randomly selected sub-ranges: $[i_1, i_1 + b/d)$, $[i_2, i_2 + b/d)$, ..., $[i_d, i_d + b/d)$. The mini-batch subgraph is then induced from the union of nodes contained in these sub-ranges. We illustrate the impact of this decomposition with an example. Suppose the batch size is $b = 6000$, the number of decompositions is $d = 2$, and we analyze the receptive field of the 10000th node $\hat{n}$ in the training graph. Without decomposition, mini-batches are formed from a single range $[i, i + 6000)$. If $\hat{n}$ appears in a mini-batch, the starting index i must lie in $[4001, 10000]$, which restricts the neighbors of $\hat{n}$ to the range $[4001, 16000)$. By contrast, when applying the decomposition strategy, mini-batches are constructed from two sub-ranges such as $[i_1, i_1 + 3000)$ and $[i_2, i_2 + 3000)$. If $\hat{n}$ belongs to the first sub-range, the index i_2 may take any value, allowing the neighbors of $\hat{n}$ to be drawn from any node in the training graph. This effectively yields a full receptive field for $\hat{n}$. Through decomposition, the receptive field of each node is expanded without increasing the overall batch size. Equivalently, for an edge (u, v) in the sampled subgraph, since u and v may originate from different sub-ranges, the value of $|u - v|$ is no longer constrained, making any edge in the graph eligible for sampling. As a result, the small receptive field issue is resolved.

All other aspects of the training procedure remain identical to those of the vanilla FOS-GNN. In the following sections, unless stated otherwise, we use FOS-GNN to refer to the model equipped with the decomposition enhancement by default.

Guidelines on Setting the Important Hyperparameters

Although the definition of FOS-GNN is conceptually simple, its training process requires careful tuning of hyperparameters. In particular, the sampled subgraph size b and the number of decompositions d are two critical and distinctive hyperparameters of FOS-GNN. By imposing appropriate lower and upper bounds on b and d, we

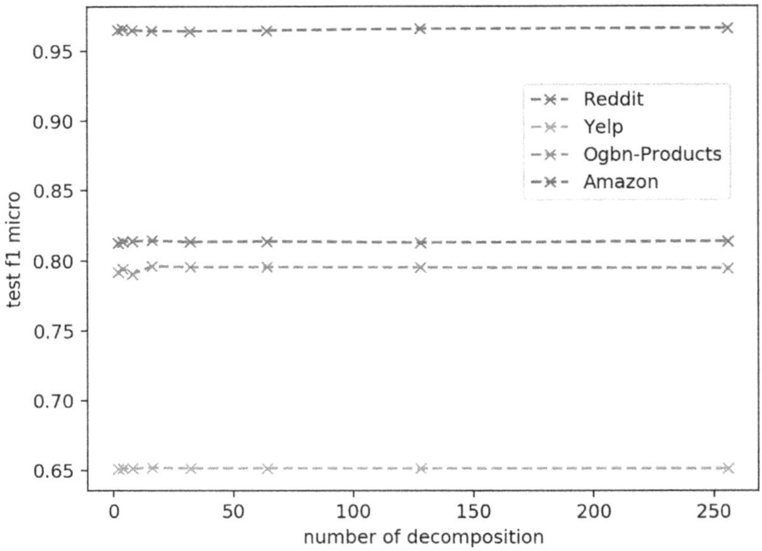

Fig. 3.3 With different numbers of decomposition (2, 4, 8, 16, 32, 64, 128, 256), the F1-micro of FOS-SANIT with decomposition on four datasets. All the other hyperparameters are fixed

can ensure the effectiveness of the FOS sampler. This section provides practical guidelines for estimating these bounds (Tables 3.3, 3.4, 3.5 and 3.6).

We first consider the mini-batch size b, which determines the average degree of a sampled subgraph and thereby affects model performance. If a mini-batch is overly sparse, the model will struggle to aggregate and propagate information, leading to poor convergence and degraded performance of FOS-GNN. Therefore, we derive a lower bound on b to guarantee that the sampled subgraph is sufficiently dense on average. Since the input graph is randomly permuted at the node level prior to training, the b vertices in a mini-batch can be regarded as uniformly sampled without replacement from all vertices. In this case, the number of edges in the sampled subgraph $\mathcal{G}_s$ and its average degree can be computed as follows:

$$\begin{aligned} |\mathcal{E}_s| &= \frac{b^2}{N^2} \cdot |\mathcal{E}|, \\ \text{average degree of } \mathcal{G}_s &= \frac{|\mathcal{E}_s|}{|\mathcal{V}_s|} = \frac{|\mathcal{E}| \cdot b}{N^2}. \end{aligned} \tag{3.3}$$

To enable effective information flow, each vertex in a mini-batch should, on average, have at least one neighbor, implying that the average degree of $\mathcal{G}_s$ should be no smaller than 1. This condition serves as a minimal requirement to reduce the hyperparameter search space. As reported in the last column of Table 3.6, in practice we typically ensure that each node has more than one neighbor. Based on this criterion, we can derive a lower bound on the batch size b as follows:

$$b_{\min} = \frac{N^2}{|\mathcal{E}|}. \tag{3.4}$$

In addition, increasing the batch size leads to higher GPU memory consumption during training. Hence, an upper bound on b can be determined according to the available GPU memory budget. Since each mini-batch contains a fixed number of vertices and incurs nearly identical computational cost, GPU memory usage remains stable across training iterations. As a result, the maximum feasible batch size $b_{\max}$ can be identified through a small number of trial runs. Beyond memory constraints, excessively large batch sizes may also result in suboptimal learning behavior, as observed in many deep learning applications. Therefore, following common practice, we further refine the choice of b via grid search.

Next, we examine the role of the number of decompositions d, which controls the randomness of vertex selection in a mini-batch and directly affects both the effectiveness and efficiency of FOS-GNN.

On one hand, increasing d enhances the diversity of nodes within a mini-batch, which can potentially improve model accuracy. We conducted a case study to investigate the impact of d on the accuracy of FOS-SAINT across four datasets. As illustrated in Fig. 3.3, the model accuracy generally stabilizes when $d \geq 8$. Overall, the performance of FOS-GNN with decomposition is relatively insensitive to the choice of d, because any pair of nodes can appear in the same mini-batch as long as $d \geq 2$, thereby satisfying the full receptive field requirement discussed in Sect. 3.2.2.

On the other hand, a large value of d causes the feature component, constituting the dominant portion of data transfer in a mini-batch, to be split into many small blocks of node features. This necessitates either transferring multiple small feature blocks to the GPU and concatenating them before training, or first aggregating them on the CPU and then transferring the combined tensor to the GPU. The latter approach incurs substantial CPU I/O overhead, similar to prior methods, while the former introduces inefficiencies in GPU data transfer. Both scenarios can significantly slow down training. In practice, most modern GNN frameworks, including DGL and PyTorch Geometric, rely on direct memory access (DMA) engines to move data from host memory to GPU. However, DMA engines are optimized primarily for transferring **large** contiguous memory blocks. Transferring small blocks severely degrades DMA performance. As shown in prior work [20], achieving 90% of the peak PCIe 3.0×16 bandwidth requires a data block size of at least 256KB. Depending on the PCIe bandwidth and the desired utilization, users may consult prior studies or benchmark their systems to identify the minimum effective block size. By enforcing that the size of each decomposition exceeds this minimum block size, we can derive an upper bound on the number of decompositions d. Let B denote the minimum block size required by the DMA engine, and let k represent the size of a single vertex's feature vector. Then the upper bound on d can be expressed as:

$$d_{\max} = \frac{b \cdot k}{B}. \tag{3.5}$$

With the above guidelines, we can substantially reduce the search space of b and d for FOS-GNN on different graphs, while simultaneously guaranteeing both effectiveness and efficiency.

We demonstrate the hyperparameter selection process using the Reddit dataset together with the FOS-SAINT model. We first determine the batch size b. Under the inductive setting, the Reddit training graph contains 153431 vertices and 52284760 edges. According to the derived lower bound, we obtain $b_{\min} = \frac{153431^2}{52284760} = 450.2$. Empirically, after several trial runs, we observe that the largest feasible batch size $b_{\max}$ exceeds 100,000, indicating that GPU memory does not impose a restrictive upper bound in this case. Consequently, the only constraint on the hyperparameter search space is $b \geq 451$.

Next, we determine the number of decompositions d. Our experimental platform is equipped with PCIe 3.0×16, and we aim to achieve more than 90% bandwidth utilization. Based on prior work [20], this requires a minimum data transfer block size of 256KB. In the Reddit dataset, each vertex is associated with 602 float32 features, resulting in a storage requirement of 2408 bytes per vertex. Therefore, for a given batch size b in the search space, the corresponding upper bound on the number of decompositions is $d^1_{\max} = \frac{2408}{256 \times 1024} \cdot b$. As discussed earlier with reference to Fig. 3.3, searching for d within the interval [2, 8] is sufficient to ensure model effectiveness. Accordingly, we select the final search range for d as $[2, \min(d^1_{\max}, 8)]$.

Theoretical Analysis of FOSSampler

Efficiency Analysis on the Reduction of Memory I/Os. As discussed in Sect. 3.1 and Sect. 3.2.2, when node features are stored in a non-contiguous manner, the DMA engine incurs additional memory I/Os during data transfer. In this part, we analyze the reduction in memory I/Os for node features when replacing topology-oriented sampling with feature-oriented sampling.

Proposition 3.1 *The FOS sampler can reduce the number of memory I/Os for node features by a factor of* $\frac{2X+1}{d}$ *compared with topology-oriented sampling methods, where* X *denotes the batch size and* d *represents the number of decompositions used by the FOS sampler.*

Proof Topology-oriented sampling approaches first perform index selection to gather all required entries from the original feature tensor and then write these entries into a contiguous buffer. Subsequently, the DMA engine issues a read transaction on this prepared buffer. As a result, this process incurs a total of $2X + 1$ memory I/Os. In contrast, the FOS sampler directly accesses d contiguous blocks of node features by exploiting the base address and offsets of the feature tensor. The DMA engine then performs d read transactions corresponding to these blocks, leading to only d memory I/Os in total. □

Effectiveness Analysis on the Utilization of Input Features. Under the transductive learning setting, the FOS sampler is able to exploit a larger portion of information than topology-oriented sampling methods.

Table 3.2 Dataset statistics (In Classes, "m" means multi-label classification and "s" means single-label classification)

Dataset	Nodes	Edges	Degree	Feature	Classes	Train/Val/Test
Reddit	232,965	115M	497	602	41 (s)	0.66/0.10/0.24
Yelp	716,847	14M	19	300	100 (m)	0.75/0.10/0.15
Ogbn-products	2,449,029	124M	50	100	47 (s)	0.08/0.02/0.90
Amazon	1,598,960	264M	168	200	107 (m)	0.85/0.05/0.10

Proposition 3.2 *The FOS sampler can utilize more node features than topology-oriented samplers by times of* $\frac{|G|}{|G_{T,L}|}$. *Here, G is the input graph, T is the training set, L denotes the model depth, and* $G_{T,L}$ *denotes L-hop neighborhood of T.*

Proof From the definition of topology-oriented samplers, we know that mini-batches are generated by sampling from $G_{T,L}$. For instance, the full neighbor sampler includes all nodes in $G_{T,L}$ but only retains edges along traversal paths originating from T. The SAGE sampler further masks a subset of nodes and edges based on the full neighbor sampler, while the random walk sampler restricts nodes and edges to those induced by L-step random walks starting from T. Consequently, the maximum number of node features that can be utilized is limited to $|G_{T,L}|$. In contrast, the FOS sampler considers all input node features as potential candidates and is therefore able to leverage up to $|G|$ node features. □

Beyond the advantages discussed earlier, the FOS sampler exhibits an additional desirable property under the inductive learning setting. Specifically, FOS provides unbiased access to all input node features, whereas topology-oriented samplers tend to favor high-degree nodes. Such bias in the sampled data can make it more difficult for GNNs to faithfully learn representations from the input graph. Let $A \in \mathbb{R}^{|T|\times|T|}$ denote the adjacency matrix of the training graph, and let $V \in \mathbb{R}^{|T|}$ represent the access probability vector over all training nodes. By definition, each training node is accessed uniformly by the FOS sampler, yielding $V_i^{\text{FOS}} = \frac{1}{|T|}$. In contrast, for the full neighbor sampler, the access probability is $V_i^{\text{Full}} = \frac{\deg(i)}{\|A\|_1}$. For the SAGE sampler, each node samples its neighbors uniformly, leading to $V_i^{\text{SAGE}} = \sum_{j\in N(i)} \frac{1}{\deg(i)}$. For the random walk sampler, the transition probability is defined as $p_{u,v} = \frac{1}{\deg(u)}$ if $v \in N(u)$ and 0 otherwise, which results in $V_i^{\text{RW}} = \frac{\deg(i)}{2m}$, where m denotes the total number of edges. Consequently, all three of these samplers exhibit bias toward nodes with higher degrees. As an illustration, under the inductive learning setting on the Ogbn-Products dataset, the top 10% highest-degree nodes account for 10.0%, 18.4%, 25.1%, and 22.8% of total node accesses for the four samplers, respectively.

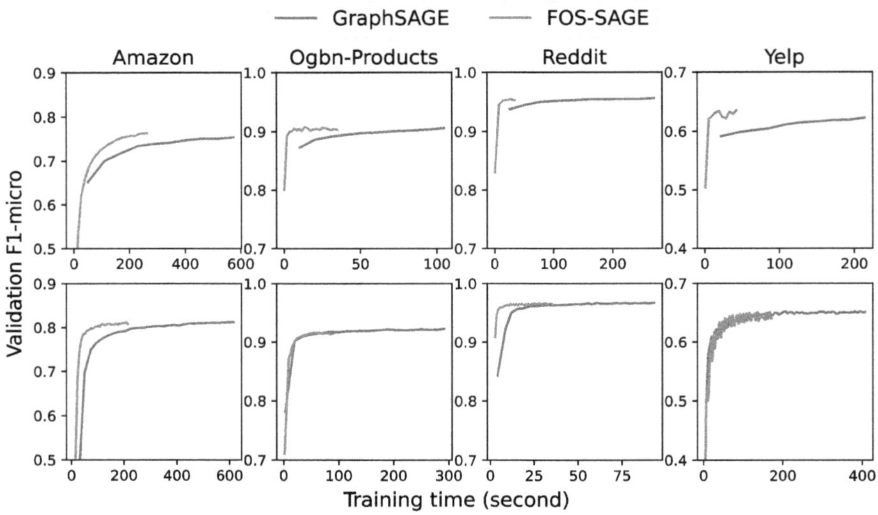

Fig. 3.4 Convergence curves of FOS-GNNs and baselines under inductive learning setting

3.2.3 *Experiments*

To assess both the effectiveness and efficiency of the FOS sampler, we carried out comprehensive experimental evaluations on four large-scale datasets under multiple learning scenarios. In particular, we first compare FOS-SAGE and FOS-SAINT against their respective baseline methods under the inductive learning setting in Sect. 3.2.3. We then further evaluate, in Sect. 3.2.3, the accuracy and convergence behavior of GCN and FOS-GCN under the transductive learning setting.

Setup. All experiments are conducted in the supervised learning setting. We evaluate FOS-GNNs and baseline methods on four representative tasks: (1) predicting post communities given user comments (Reddit); (2) predicting business categories based on customer reviews and social connections (Yelp); (3) classifying product categories using an Amazon co-purchasing network released by the Open Graph Benchmark (Ogbn-Products) [2]; and (4) predicting product categories from customer reviews and interaction data (Amazon). As summarized in Table 3.2, all four datasets are large in scale, publicly available, and commonly adopted in prior studies [1, 3, 7]. Although Ogbn-Products is primarily used in transductive learning settings and has a relatively small training ratio of 8%, we additionally evaluate it under the inductive setting in Sect. 3.2.3 to better examine the generalization capability of FOS-GNN and the baseline approaches.

To ensure a fair comparison and avoid the influence of dataset-specific engineering tricks, we adopt the official implementations provided in DGL's public examples for all baseline methods.[1] Building upon these implementations, we develop three variants of FOS-GNN as described in Sect. 3.2.2 and release them as open-source software.[2] In addition, we adapt the full neighbor sampler to the original GCN model [1], enabling its training on large-scale graphs under constrained GPU memory.

For all methods, we conduct a grid search over the hyperparameter space defined as follows:

- Hidden dimension: $\{128, 256, 512\}$
- Dropout: $\{0.0, 0.1, 0.2, 0.3\}$
- Learning rate: $\{0.1, 0.01, 0.001, 0.0001\}$

We adopt the experimental configurations from prior work [3] for the hidden dimensions on each dataset, setting them to 128 for Reddit, and 512 for Yelp, Ogbn-Products, and Amazon. For the number of seed nodes used in vanilla mini-batch GCN and GraphSAGE, we conduct a grid search over {256, 512, 1024, 2048}, and observe that the optimal performance is achieved with 512 seeds for GraphSAGE and 1024 seeds for GCN. In the case of GraphSAGE, we employ the mean aggregator with fanout sizes of 10 and 25. For GraphSAINT, we utilize its random walk sampler. On the Reddit, Yelp, and Amazon datasets, we adopt the best-performing hyperparameters reported in the original GraphSAINT study. For Ogbn-Products, we additionally explore the num_roots parameter of the random walk sampler over $\{1000, 1500, 2000, 3000\}$ as part of the hyperparameter search. All methods are optimized using Adam [21]. Model accuracy is evaluated on the test set using the F1-micro metric. For all approaches, the number of GCN layers is fixed to 2.

Each method is trained for a fixed number of epochs determined by convergence behavior. During training, we checkpoint the model parameters that achieve the highest F1 score on the validation set. The selected model is then evaluated on the test set, and the corresponding test accuracy is reported.

All experiments are executed on a single machine equipped with an Intel(R) Xeon(R) Gold 6240 CPU running at 2.60GHz, an NVIDIA GeForce RTX 2080 Ti GPU with 11GB of HBM memory connected via PCIe 3.0×16, and 256GB of DDR4 system memory. The implementation is written in Python 3.7.11. We use DGL 0.7.1 and PyTorch 1.8.0 with CUDA 10.2 and cuDNN 8.0.3 for GPU training. During training, OMP_NUM_THREADS is set to 20, and no additional dataloader worker processes are employed for any of the methods.

Inductive Accuracy and Speed Comparison

The accuracy and convergence behaviors of all evaluated methods are reported in Table 3.3 and Fig. 3.4. The training time illustrated in Fig. 3.4 excludes the overhead of data loading, preprocessing, validation evaluation, and model checkpointing. Detailed training configurations are summarized in Table 3.6. The results demonstrate

---

[1] URL: https://github.com/dmlc/dgl/tree/master/examples Commit: a04a8d.

[2] https://github.com/initzhang/FOSGNN.

Table 3.3 Comparison of F1-micro score and convergence time between FOS-GNNs and baselines under inductive learning setting

	Reddit		Yelp		Ogbn-products		Amazon	
	F1-micro	Time(s)	F1-micro	Time(s)	F1-micro	Time(s)	F1-micro	Time(s)
GraphSAGE	0.960±0.001	270.8 (7.9×)	0.624±0.002	213.8 (5.1×)	0.754±0.002	104.6 (3.0×)	0.755±0.001	573.4 (2.2×)
FOS-SAGE	0.959±0.001	34.1	0.636±0.001	42.2	0.749±0.005	34.9	0.763±0.001	262.6
GraphSAINT	0.966±0.001	204.3 (2.8×)	0.653±0.002	473.7 (2.6×)	0.787±0.001	318.7 (2.7×)	0.814±0.001	1447.1 (3.5×)
FOS-SAINT	0.967±0.001	72.9	0.652±0.001	183.5	0.795±0.003	117.3	0.814±0.002	413.2

The convergence time includes training time and preprocessing time

Table 3.4 Top 10 most time-consuming operators when training FOS-SAINT on the Amazon dataset

Operator name	Wall duration (ms)	Self time (ms)	Average wall duration (ms)	Occurrences
Optimizer.step#Adam.step	41.898	12.681	10.475	4
aten::index_select	38.594	37.923	0.568	68
cudaLaunchKernel	36.983	36.983	0.012	2977
aten::to	27.850	2.740	0.043	653
aten::mul	25.547	11.530	0.052	438
aten::copy_	24.217	0.898	0.378	64
cudaStreamSynchronize	15.467	15.467	0.204	76
aten::sum	14.318	7.684	0.083	172
GSpMM	13.532	8.479	0.483	28
aten::add_	11.886	6.754	0.028	419

We set the batch size to 7847, which is the average number of nodes per mini-batch under settings in Table 3.1. The first 4 iterations are warm-up stage and we measure the accumulated training time of the following 4 iterations with PyTorch Profiler [18]

that both variants of FOS-GNN achieve accuracy comparable to their corresponding baseline methods across all datasets, while substantially accelerating the training process. To further investigate the sources of this speedup, we conduct four targeted case studies that analyze different aspects of the FOS sampler.

First, we employ the PyTorch Profiler to examine how FOS-GNNs benefit from reducing random accesses to node features. Specifically, we profile the training process of FOS-SAINT under the same experimental settings as those used in Table 3.1. The profiling results are presented in Table 3.4. By removing random access to node features, the wall-clock time of the *aten*::**index_select** operator is reduced from 214 to 39 ms. The operator's execution time does not drop to zero because it is still invoked during the construction of mini-batch adjacency matrices.

Second, Table 3.5 demonstrates that FOS-SAINT is able to reduce both preprocessing overhead and overall training time when compared with GraphSAINT. During training, GraphSAINT relies on a random walk sampler to generate subgraphs from randomly scattered nodes, which leads to expensive random accesses to the corresponding data, as evidenced in Table 3.1. In contrast, FOS-SAINT removes a substantial portion of these random accesses, as shown in Table 3.4, thereby significantly accelerating the training process. With respect to preprocessing, GraphSAINT incurs heavy computational cost to estimate normalization coefficients, whereas FOS-SAINT can compute its normalization terms with minimal overhead. Although preprocessing might be viewed as a one-time cost, this assumption only holds when hyperparameters are already fixed. In practice, preprocessing-related hyperparameters (e.g., number of roots and random walk length in GraphSAINT) have a strong impact on model accuracy [3], and lengthy preprocessing can substantially complicate hyperparameter tuning and model training.

Table 3.5 Comparison of preprocessing time and training time for FOS-SAINT and GraphSAINT

	Reddit		Yelp		Ogbn-products		Amazon	
	Pre-processing	Training	Pre-processing	Training	Pre-processing	Training	Pre-processing	Training
GraphSAINT	110.8 (2.9×)	93.5 (2.7×)	67.2 (5.5×)	406.5 (2.4×)	27.2 (0.7×)	291.5 (3.8×)	831.7 (4.2×)	615.4 (2.9×)
FOS-SAINT	37.7	35.2	12.2	171.3	39.7	77.6	197.9	215.3

(unit: second)

Table 3.6 Training configuration for FOS-GNNs in Tables 3.3 and 3.8

Method	Datasets	Dim.	Learning rate	Dropout	Batch size	Num. Decomp.	Avg. Degree
FOS-GCN	Ogbn-Products	512	0.001	0.2	30,000	8	2
FOS-SAGE	Reddit	128	0.01	0.3	20,000	8	43
	Yelp	512	0.01	0.2	10,000	8	1.3
	Ogbn-Products	512	0.01	0.2	30,000	8	8
	Amazon	512	0.001	0.2	20,000	8	3
FOS-SAINT	Reddit	128	0.01	0.2	9000	4	21
	Yelp	512	0.1	0.1	29,000	8	2
	Ogbn-Products	512	0.001	0.3	36,000	4	9
	Amazon	512	0.1	0.1	40,000	4	5

Table 3.7 shows that FOS-SAINT achieves better GPU utilization and alleviates CPU contention during training. Random feature access triggers intensive CPU I/O operations, which can starve the GPU and limit throughput. By removing this bottleneck, FOS-SAINT shifts the workload away from the CPU and achieves up to 3.8× speedup in training and up to 5.5× speedup in preprocessing compared with GraphSAINT. Furthermore, the FOS sampler exhibits superior scalability with respect to batch size compared with the SAGE sampler, not to mention the full neighbor sampler. For the latter two approaches, batch size grows exponentially as model depth increases, which is inherent to their sampling mechanisms. By contrast, the batch size of the FOS sampler is only indirectly affected by model performance. We conduct a grid search over the batch size parameter for the FOS sampler, ranging from 10,000 to 50,000 with a step size of 5000, and observe that relatively small batch sizes are sufficient to reach optimal performance. As illustrated in Fig. 3.5, the average batch size produced by the FOS sampler remains consistently smaller and increases more slowly than that of the SAGE sampler as the model depth grows.

Table 3.7 Comparison of CPU usage and GPU utilization during training of GraphSAINT and FOS-SAINT.

	Reddit		Yelp		Ogbn-Products		Amazon	
	CPU (%)	GPU (%)	CPU (%)	GPU (%)	CPU (%)	GPU (%)	CPU (%)	GPU (%)
GraphSAINT	373	17	1044	46	594	56	967	24
FOS-SAINT	100	25	100	85	100	75	100	68

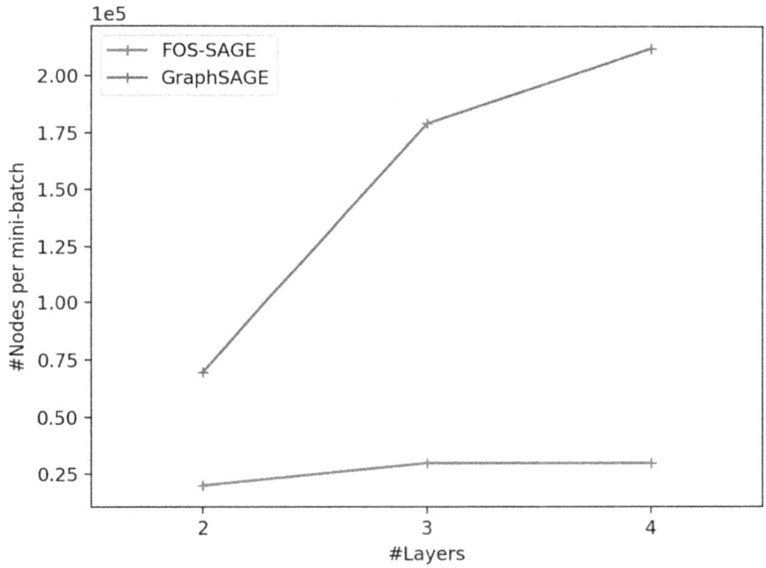

Fig. 3.5 Average batch size with respect to model depth on Reddit. For GraphSAGE, the fanout of each layer is fixed as 15 and the number of seeds equals 512. For FOS-SAGE, we perform grid search to determine the batch size so that the two methods have similar performance

Finally, we observe that the advantages of adopting feature-oriented sampling become more pronounced on graphs that are both larger in scale and denser in connectivity, such as the Reddit and Amazon datasets. This is because topology-oriented sampling methods on such graphs typically incur substantial overhead when constructing mini-batches, particularly due to the high cost of randomly accessing node features.

Transductive Accuracy and Speed Comparison

The accuracy and convergence behaviors of GCN and FOS-GCN are reported in Table 3.8 and Fig. 3.6. Consistent with the observations in Sect. 3.2.3, FOS-GCN significantly outperforms vanilla mini-batch GCN in terms of training efficiency,

Table 3.8 Comparison of F1-micro score and convergence time between GCN and FOS-GCN under transductive learning setting

	F1-micro	Time(s)
GCN	0.747 ± 0.008	1322.2 (4.5×)
FOS-GCN	0.793 ± 0.003	290.6

The convergence time includes training time and preprocessing time

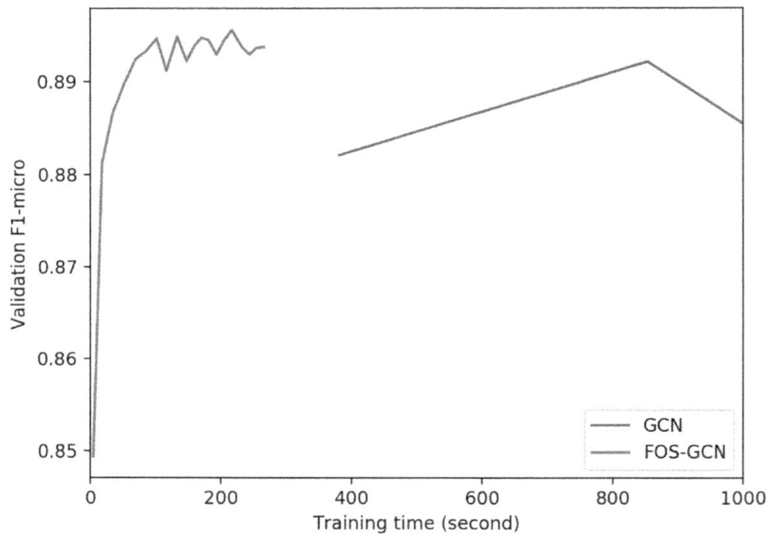

Fig. 3.6 Convergence curves of GCN and FOS-GCN on Ogbn-Products under transductive learning setting

achieving substantially shorter training time. These experimental results further confirm the advantages of the FOS sampler discussed in Sect. 3.2.2. Under the transductive learning setting, the FOS sampler is able to exploit a larger portion of node features than the full neighbor sampler employed by GCN. Specifically, the full neighbor sampler restricts information usage to the L-hop neighborhood of all training nodes, where L denotes the model depth. In contrast, the FOS sampler accesses all input node features in a uniform manner, thereby incorporating the complete set of available information. On the Ogbn-Products dataset, FOS-GCN can utilize 100% of node features with uniform access probability, whereas vanilla mini-batch GCN only covers 94% of node features and does so with a biased sampling distribution. This difference provides a clear explanation for the observed accuracy gains of FOS-GCN over standard GCN.

3.3 Dual-Cache Training

3.3.1 Drawbacks of Single-Cache Systems

As noted earlier, the time spent preparing mini-batches constitutes the major performance bottleneck in mini-batch GNN training. To reduce the significant cost incurred during the construction of each batch, substantial research has focused on accelerating Adj-Sampling and Nfeat-Selecting. Several studies [12, 13] advocate the use of unified virtual addressing (UVA) to streamline the access and manipulation of irregular graph structures. By improving the utilization of PCIe bandwidth during data transfers to the GPU, UVA enables large-scale graph structures to remain in host memory while still supporting efficient Adj-Sampling and Nfeat-Selecting on the GPU. Nevertheless, the UVA approach overlooks the inherent locality present in graph datasets, resulting in repeated movement of frequently used data and thereby introducing avoidable overhead.

Another group of studies [9, 11, 12, 14] introduces cache-based systems, which speed up mini-batch preparation by exploiting both the locality patterns in graph workloads and the typically underused GPU memory. On one side, graph data accessed during training demonstrates pronounced locality. For instance, when running GraphSAGE on the Ogbn-Papers100M dataset, about 95% of all feature lookups fall within the hottest 2GB of features.[3] On the other side, as reported in [9], up to 90% of GPU memory can remain idle throughout training. Leveraging these observations, prior cache-oriented approaches accelerate node feature retrieval by storing the most frequently accessed features in unused GPU memory (Nfeat-Cache), thereby decreasing CPU–GPU data transfers. We refer to this family of approaches as the Single-Cache systems, since they only maintain a cache for node features.

Although widely used in existing cache-oriented designs, the Single-Cache strategy implicitly depends on two key assumptions. (1) Nfeat-Selecting consistently dominates Adj-Sampling in runtime, and (2) the Nfeat-Cache can effectively exploit all available GPU memory. In practice, neither assumption is guaranteed to hold. By profiling the workloads of Adj-Sampling and Nfeat-Selecting,[4] we observe that their execution-time ratios span a wide range, from 0.39 up to 2.17. Consequently, current Single-Cache designs lack the flexibility to handle such variation, as they always focus on optimizing Nfeat-Selecting even in scenarios where Adj-Sampling becomes the dominant cost.

Moreover, our measurements reveal that a Single-Cache approach may leave a substantial portion of GPU memory ineffective, leading to waste on the order of several gigabytes. For instance, during GraphSAGE training on the Ogbn-Papers100M dataset, the Nfeat-Cache occupies only 2.5GB out of the 6GB of otherwise idle GPU memory.[5] This unused capacity suggests the possibility of adding an additional

---

[3] batch size = 8000, fanouts = 15,10,5. Statistics computed after 10 epochs of training.

[4] Details and settings will be given in Sect. 3.3.1.

[5] Details and settings will be given in Sect. 3.3.1.

Table 3.9 Comparison of DUCATI with the recent works. DCA represents the Dual-Cache Allocator. SOTA represents the state-of-the-art system to be introduced in Sect. 3.3.3

Systems	Adj		Nfeat		
	Cache	UVA	Cache	UVA	DCA
PyTorch-Direct [13]	✗	✗	✗	✓	✗
PaGraph[9]/GNNLab[11]	✗	✗	✓	✗	✗
Quiver [12]	✗	✓	✓	✗	✗
DGL [17]	✗	✓	✗	✓	✗
SOTA	✗	✓	✓	✓	✗
DUCATI	✓	✓	✓	✓	✓

cache to accelerate Adj-Sampling, whose access patterns also demonstrate locality.[6] Building on these insights, we contend that pairing the existing Nfeat-Cache with a complementary Adj-Cache is both practical and advantageous.

In this section, we introduce DUCATI, a Dual-Cache Training System designed to flexibly handle diverse workloads while more efficiently utilizing GPU memory. Beyond the conventional Nfeat-Cache, DUCATI also takes advantage of locality in the adjacency matrix by incorporating a new Adj-Cache to speed up Adj-Sampling. Constructing such a Dual-Cache training framework raises two key challenges.

First, how should we design the Adj-Cache? The methodology behind Nfeat-Cache does not directly apply because the properties and processing patterns of adjacency data differ from those of node features. Node features share a uniform length, whereas adjacency entries vary in size. Moreover, feature accesses are relatively dense during Nfeat-Selecting, while adjacency accesses are dispersed and intertwined with fine-grained operations during Adj-Sampling.

Second, how can GPU memory be jointly allocated between Adj-Cache and Nfeat-Cache? Naturally, the two caches compete for the same limited GPU memory. Expanding the budget for Adj-Cache reduces the space available to Nfeat-Cache. As a result, decreasing Adj-Sampling latency by enlarging Adj-Cache will unavoidably increase the cost of Nfeat-Selecting. Adding to the complexity, the relative workloads of Adj-Sampling and Nfeat-Selecting vary across different configurations, requiring the system to adaptively determine the optimal allocation strategy for each setting.

To overcome the first challenge, DUCATI introduces a dedicated cache design for Adj-Cache that aligns with both the data-access patterns of Adj-Sampling and the structural properties of the adjacency matrix. For the second challenge, DUCATI incorporates a workload-aware Dual-Cache Allocator, which dynamically determines the optimal division of GPU memory between Adj-Cache and Nfeat-Cache so as to minimize overall runtime. Table 3.9 provides a comparison of the distinct techniques adopted by prior approaches and by DUCATI.

DUCATI introduces only lightweight modifications to the programming interface of DGL, a widely used GNN training framework, and consistently accelerates mini-

[6] Details and settings will be given in Sect. 3.3.1.

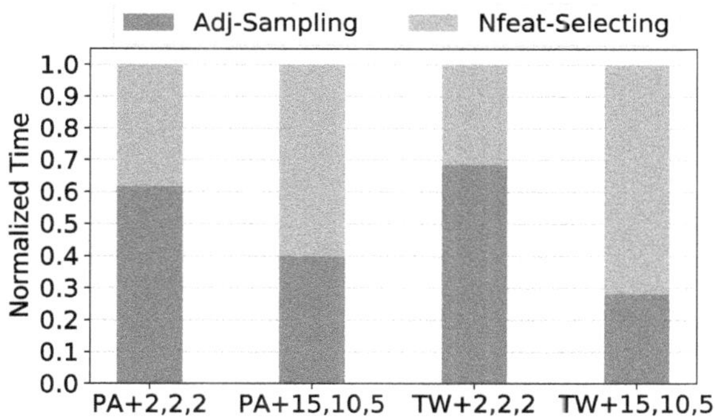

Fig. 3.7 Comparison of normalized time between Adj-Sampling and Nfeat-Selecting under different settings (datasets+fanouts). PA represents Ogbn-Papers100M and TW represents Twitter. (batch size = 8000; nfeat dimension = 128)

batch training across diverse workload scenarios. Through extensive experiments on four billion-scale datasets, we demonstrate that DUCATI delivers up to a 3.33× performance improvement (2.07× on average) over DGL, and achieves as much as a 1.54× speedup (1.32× on average) compared with the most advanced Single-Cache baselines.

Moreover, a variety of GNN training frameworks have recently been proposed [10, 12, 17, 22] to support either mini-batch or full-batch training. To assist practitioners in selecting an appropriate system, we further examine the time–accuracy trade-offs of DUCATI alongside four state-of-the-art alternatives. Our experiments indicate that, whenever the memory needed for training fits within the CPU's capacity, DUCATI consistently offers the most favorable performance for mini-batch GNN training.

Now we present empirical findings that highlight the shortcomings of the Single-Cache approaches and reveal insights that motivate the design of the Dual-Cache system.

Adjacency-matrix accesses also demonstrate locality. Our measurements show that accesses to the adjacency matrix follow locality patterns similar to those observed for node features. For instance, when training GraphSAGE on the Ogbn-Papers100M dataset, only 0.7GB out of the full 12.9GB adjacency matrix accounts for over 98% of all adjacency lookups.[7]

Adj-Sampling may dominate the runtime over Nfeat-Selecting. The relative costs of Adj-Sampling and Nfeat-Selecting vary substantially across configurations, as illustrated in Fig. 3.7. According to Amdahl's law, the performance gains of current Single-Cache approaches are inherently constrained because they focus solely on optimizing Nfeat-Selecting, even in scenarios where Adj-Sampling becomes the more time-consuming component.

Nfeat-Cache cannot fully exploit the GPU memory. As shown in Fig. 3.8, increasing the size of Nfeat-Cache yields diminishing returns. Many real-world graphs follow a power-law distribution, meaning only a small subset of node features is accessed frequently. Consequently, most of the possible performance benefit is captured using a relatively small cache. Consistent with Fig. 3.8, Nfeat-Cache gains little

[7] batch size = 8000, fanouts = 15,10,5. Statistics are calculated after 10 epochs of training.

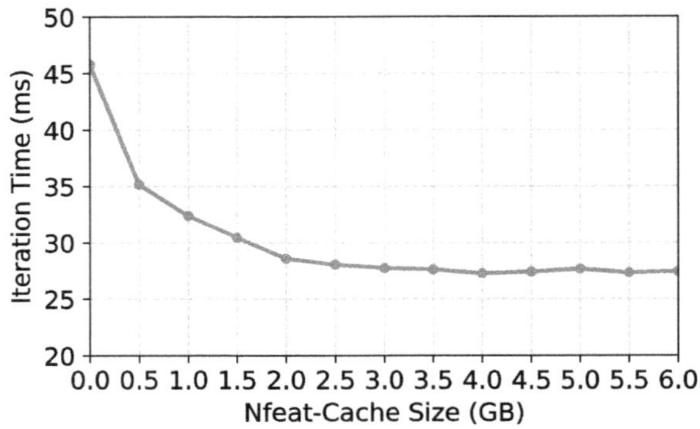

Fig. 3.8 The trend of iteration time with respect to the increase of Nfeat-Cache. We obtain the results on the Ogbn-Papers100M dataset with GraphSAGE. (fanouts = 15,10,5; batch size = 8000; nfeat dimension = 128)

improvement once its size exceeds roughly 2.5GB, leaving a considerable portion of the spare GPU memory underutilized.

Since Nfeat-Cache is unable to take advantage of all the remaining GPU memory and the relatively expensive Adj-Sampling step can be sped up through an additional Adj-Cache, these observations naturally point toward a more powerful design. We propose a Dual-Cache system that incorporates both Adj-Cache and Nfeat-Cache.

3.3.2 Designs of Two Caches and the DUCATI Framework

DUCATI is a workload-aware, cache-centric system designed for mini-batch GNN training on extremely large graphs. It leverages underutilized GPU memory to store frequently accessed portions of graph data. Unlike prior Single-Cache designs that rely solely on Nfeat-Cache, DUCATI adopts a Dual-Cache architecture incorporating both Adj-Cache and Nfeat-Cache. The added Adj-Cache helps accelerate the expensive Adj-Sampling stage and alleviates the diminishing returns typically observed in Nfeat-Cache. In addition, DUCATI features a workload-aware Dual-Cache Allocator capable of adapting to different training conditions. Under any given configuration, the allocator identifies whether Adj-Sampling or Nfeat-Selecting dominates the mini-batch generation time and then assigns more memory to the corresponding cache (Adj-Cache or Nfeat-Cache). In essence, DUCATI dynamically determines the optimal distribution of spare GPU memory between the two caches to achieve the highest possible training throughput.

Note that mini-batch preparation constitutes the dominant bottleneck in end-to-end GNN training, taking roughly 60–95% of the total runtime. Under the PA (15,10,5 + 256) configuration, our dual-cache mechanism can cut this preparation time by as much as 80%. For this reason, DUCATI is designed specifically to address the crucial challenge, i.e., the non-trivial dual-cache design within the mini-batch preparation pipeline, rather than focusing on the model-training phase of the overall workflow.

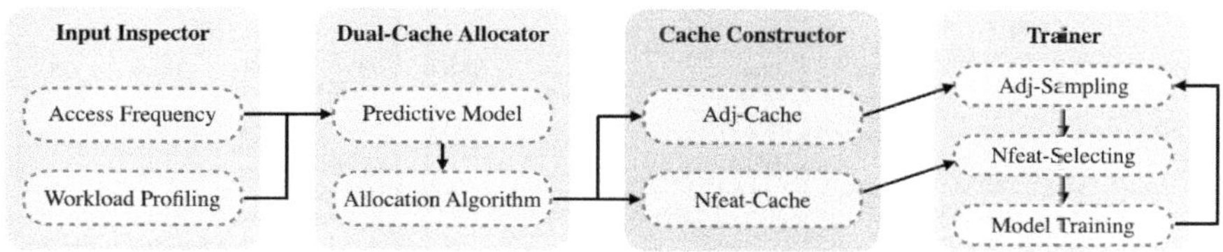

Fig. 3.9 Overview of DUCATI

An overview of DUCATI is provided in Fig. 3.9. The system first extracts key information from the inputs using the Input Inspector. Based on this extracted information, the Dual-Cache Allocator determines the optimal memory allocation strategy under the current workload to maximize throughput. Afterward, the Cache Constructor builds both Adj-Cache and Nfeat-Cache accordingly. With these two caches in place, the system proceeds to perform mini-batch training. Each of these components will be described in detail in the subsequent sections.

Figure 3.10 illustrates how DUCATI differs from existing approaches when handling Adj-Sampling and Nfeat-Selecting on massive graphs whose adjacency matrices and node features exceed GPU memory capacity. Prior UVA-based methods [12, 13] keep both adjacency data and node features in pinned host memory and rely on GPU kernels for Adj-Sampling and Nfeat-Selecting. Earlier cache-based systems [9, 11, 12, 14] store these graph components in host memory while keeping only Nfeat-Cache on the GPU; in these systems, Adj-Sampling runs on the CPU, whereas Nfeat-Selecting runs on either CPU or GPU depending on cache hits. In contrast, DUCATI stores the adjacency matrix and node features in pinned host memory, while placing both Adj-Cache and Nfeat-Cache on the GPU. DUCATI performs both Adj-Sampling and Nfeat-Selecting using GPU kernels. Cache hits are served directly from GPU memory, and cache misses fall back to UVA-based data fetching. As shown, previous approaches overlook the potential of Adj-Cache and thus miss a significant optimization opportunity. DUCATI addresses this gap by integrating Adj-Cache and combining caching with UVA for maximal performance.

Input Inspector

The Input Inspector extracts two key forms of information through a set of offline sample runs. First, it computes the access frequency for every entry, which provides an estimate of each entry's true access probability. Intuitively, entries that are both frequently accessed and small in size are more desirable to place in the cache. The strategy for selecting cache contents will be detailed in Sect. 3.3.2. Second, the Input Inspector profiles the workloads of Adj-Sampling and Nfeat-Selecting, gathering statistics on how much each of them benefits from GPU caching. These workload measurements serve as the foundation for constructing the workload-aware Dual-Cache Allocator.

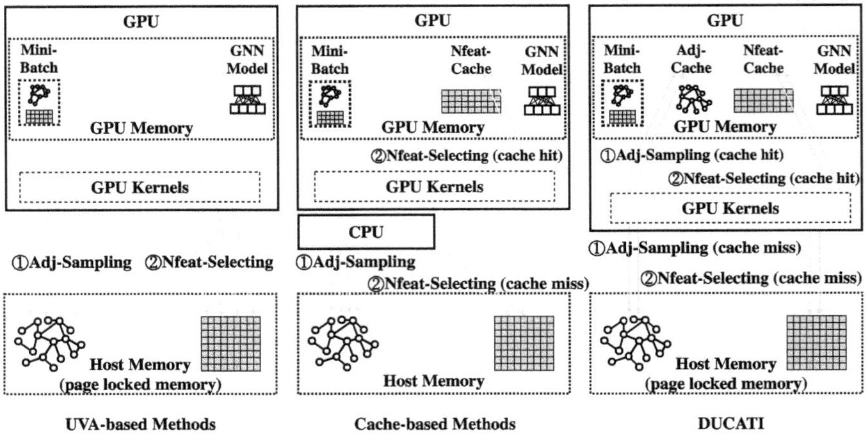

Fig. 3.10 Comparison of UVA-based methods, cache-based methods, and DUCATI on performing Adj-Sampling and Nfeat-Selecting on giant graphs

Access-frequency Information. During several offline sample runs, we track how often each entry is accessed and use this frequency to approximate the actual access probability. Prior research [11] shows that one or two epochs of sampling are sufficient to obtain accurate estimates. This frequency data will later be used in Sect. 3.3.2.
Workload-profiling Information. Different input configurations typically produce distinct runtime characteristics for Adj-Sampling and Nfeat-Selecting. Moreover, allocating the same amount of cache space to Adj-Cache or Nfeat-Cache may yield different levels of performance improvement depending on their ratios. This variability makes cache allocation challenging. To handle it, we pre-profile how each workload responds to GPU caching under the current input. Specifically, we build multiple Adj-Cache variants containing randomly selected entries and measure the corresponding Adj-Sampling runtime. We repeat the same process for Nfeat-Selecting with various Nfeat-Cache configurations. This yields a set of data points of the form (cached entries, running time). Section 3.3.2 will describe how this information is used to construct the workload-aware Dual-Cache Allocator.

Dual-Cache Allocator

The goal of the cache allocator is to evaluate how advantageous it is to cache each individual entry and then pick the ones that provide the highest overall gain. Here, the benefit of an entry refers to the expected reduction in total execution time if that entry resides in the cache. Under a single-cache design, this comparison is straightforward because all entries share identical shapes and access behaviors. As a result, estimating and ranking their benefits is simple. In contrast, our dual-cache setting introduces additional complexity because the entries are heterogeneous. DUCATI must cache

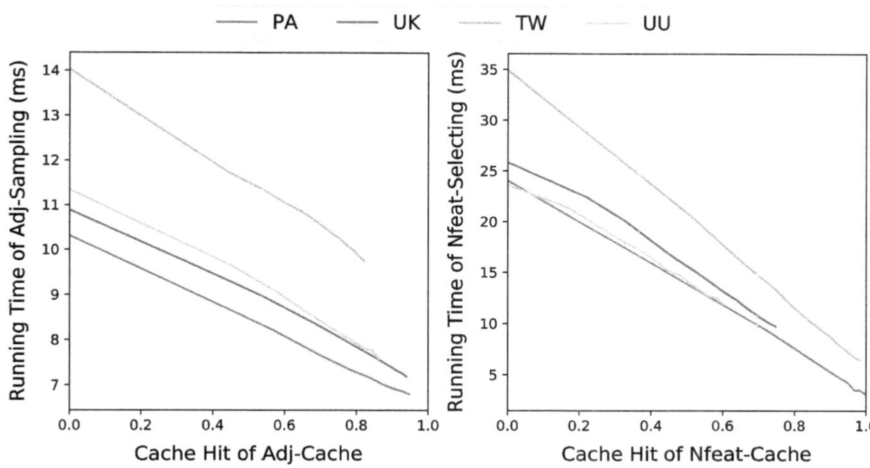

Fig. 3.11 System running time of Adj-Sampling and Nfeat-Selecting with respect to the cache hit of Adj-Cache and Nfeat-Cache

both adjacency lists and node features, and these two categories differ significantly in both data organization and runtime usage. Adjacency lists vary in length and are accessed in a fine-grained, scattered manner, whereas node features have uniform size and are retrieved once per mini-batch. Such fundamental heterogeneity complicates any direct comparison of caching benefits across the two entry types.

Moreover, several external factors further increase the difficulty of analyzing cache allocation. On the hardware side, PCIe characteristics (e.g., version, lane count, and link utilization) directly affect host-to-device transfer bandwidth. On the software side, different datasets and training configurations produce dynamic workloads for Adj-Sampling and Nfeat-Selecting. These variations cause the training speedups from the same cache budget to fluctuate widely, making it impractical to derive a clean analytical formulation solely from GNN execution semantics and graph properties.

To handle these challenges, our approach introduces a benefit prediction model, as detailed in Sect. 3.3.2. In brief, this model learns how all relevant factors influence performance through profiling and then estimates the benefit of caching any specific entry, regardless of its type. Once these benefits are available, we can directly compare all candidate entries and select those offering the highest payoff. This effectively converts the dual-cache allocation task into a knapsack-style optimization problem, for which we design a tailored algorithm to compute an effective allocation plan.

The Benefit Prediction Model

In DUCATI, because two distinct categories of data can be cached, we must devise a unified criterion for quantifying the benefit of caching any individual entry. Our objective is to construct a predictive model that connects an entry's theoretical characteristics with the empirical reduction in end-to-end execution time when that entry is placed in the cache. A key empirical observation enables us to build such a model in a concise yet effective manner. As illustrated in Fig. 3.11, we measure the execution time of Adj-Sampling and Nfeat-Selecting under multiple randomly generated Adj-Cache and Nfeat-Cache configurations of varying sizes. From these measurements, two patterns clearly emerge: (1) for each workload, the execution time exhibits an almost perfectly (inverse) linear dependence on the cache hit rate of its corresponding cache; and (2) the slope of this linear trend differs across tasks and datasets.

The near-linear relationship between cache hit rate and runtime arises because CPU–GPU data transfer time scales proportionally with the amount of data being moved, a behavior typical in systems where the two processors are connected via PCIe links [23]. This linear trend indicates that the performance gain from caching a particular entry is directly tied to its contribution to the total cache hit. Consequently, for entries within the same category, we can compare their usefulness through the individual hit metric, defined as the ratio between an entry's access frequency and the total access frequency of all entries of that category. Meanwhile, the slope differences reflect the varying strengths with which adjacency-list caching and feature caching accelerate the system. Therefore, when comparing entries across the two types, we combine the individual hit with the slope information to obtain a unified measure of benefit.

Definition 3.1 (The Benefit Prediction Model) Given the access frequency of each entry e_i as $f(e_i)$, the slope information for each type $type(e_i)$ as slope(typeof(e_i)). The benefit prediction model gives the benefit of caching each entry e_i as:

$$
\begin{aligned}
F(e_i) &= \text{slope}(\text{type of}(e_i)) \cdot h(e_i), \\
h(e_i) &= \frac{f(e_i)}{\sum_{j \in \text{type of}(e_i)} f(e_j)}.
\end{aligned}
\tag{3.6}
$$

Here $h(e_i)$ represents the individual hit of e_i.

The access frequency required to compute this value is directly provided by the Input Inspector. The slope information is also obtained from the second type of statistics collected by the Input Inspector. Concretely, we first compute the individual hit for every cached entry and convert the profiling results into a set of (cache hit, running time) pairs. We then apply a linear regression to these points to extract the slope. We adopt a linear model—instead of a polynomial or more complex model—because it aligns naturally with the underlying cache behavior: the execution time of Adj-Sampling or Nfeat-Selecting scales proportionally with the amount of data transferred, and as the cache hit increases linearly, the transferred volume decreases linearly as well.

It is important to emphasize that this benefit prediction model is inherently workload-aware. When the workloads of Adj-Sampling and Nfeat-Selecting shift due to variations in the input, the distribution of individual hit values also changes. As a result, the model F continually reflects the current access patterns and can accurately estimate the benefit of caching each entry under the present workload.

The Dual-Cache Allocation Problem
We denote the total cache budget, the cache budget of Adj-Cache, and the budget of Nfeat-Cache as B, B_{adj}, and $B_{n\text{feat}}$, respectively. Following the convention of previous works [9, 11], we estimate B with several sample runs, as the GPU memory consumption remains similar across training iterations. Using the benefit prediction model introduced in Sect. 3.3.2, we obtain the benefit of caching each entry in terms of its contribution to reducing the overall system running time. The cost of caching an entry corresponds to the size of its adjacency list or node feature.

Definition 3.2 (The Dual-Cache Allocation Problem) Given $2N$ entries (N adjacency lists and N node features), the benefit $F(e_i)$ (given by the benefit prediction model) and the cost $S(e_i)$ (in terms of size) of caching each entry e_i, and a total cache budget B. The dual-cache allocation problem is defined as follows:

$$\begin{aligned} &\text{maximize} \sum_{i=1}^{2N} F(e_i)x_i, \\ &\text{subject to} \sum_{i=1}^{2N} S(e_i)x_i \le B \text{ and } x_i \in [0, 1]. \end{aligned} \tag{3.7}$$

From this definition, the dual-cache allocation task becomes a divisible variant of the classic knapsack problem. To address it, we design Algorithm 3, which efficiently solves the allocation and can be proven to achieve the optimal solution.

The complexity of Algorithm 3 is $O(N\log N)$, which is caused by the sorting of all entries. The execution time of the algorithm can be amortized during the training as will be discussed in Sect. 3.3.3.

Proposition 3.3 *The output of Algorithm 1 is an optimal solution to the dual-cache allocation problem.*

Proof of Proposition 3.3 We prove that Algorithm 3 is always at least as better as the optimal solution during each iteration of the algorithm. Specifically, we investigate the intermediate result after step k of the for loop in Algorithm 3. We denote the set of entries chosen by Algorithm 3 as $E = (e'_1, \ldots, e'_k)$, denote the benefit as $X_1 = \sum_{j=1}^{k} F(e'_j)$, and denote the cost as $Y_1 = \sum_{j=1}^{k} S(e'_j)$. We denote the set of entries chosen by the optimal algorithm with a maximum cost of Y_1 as $O = (o_1, o_2, \ldots, o_p)$. The benefit obtained by the optimal algorithm is $X_2 = \sum_{t=1}^{p} F(o_t)$. We prove that $X_1 \ge X_2$ as follows:

- If O is identical to E, then Algorithm 3 produces the same result as the optimal algorithm. Thus $X_1 = X_2$
- If O is different from E, then either O is a proper subset of E, or O substitutes some elements in E with entries of lower benefit density (since E contains k entries with the largest benefit density by definition of Algorithm 3). In both cases, we have $X_1 > X_2$.

Algorithm 3 The Dual-Cache Allocation Algorithm.

Require: $2N$ entries (N adjacency lists and N node features); total cache budget B; the benefit $F(e_i)$ and the cost $S(e_i)$ of caching each entry e_i

Ensure: B_{adj}, B_{nfeat}, and x_i

$B_{adj} \leftarrow 0$, $B_{nfeat} \leftarrow 0$, $costs \leftarrow 0$, $x_i \leftarrow 0$
Sort all entries in the descending order of $F(e_i)/S(e_i)$. We denote the sorted entries as e'_j, and the map from the sorted indices to the original indices as M
for $j = 1, 2, \ldots, 2N$ **do**
    if $costs + S(e'_j) \leq B$ **then**
        ▷ Cache the current entry.
        $x_{M[j]} \leftarrow 1$, $costs \leftarrow costs + S(e'_j)$
        if type(e'_j) = adj **then**
            $B_{adj} \leftarrow B_{adj} + S(e'_j)$
        else
            $B_{nfeat} \leftarrow B_{nfeat} + S(e'_j)$
        end if
    else if $costs < B$ **then**
        ▷ Cache a part of the current entry.
        $x_{M[j]} \leftarrow \frac{B-costs}{S(e'_j)}$, $costs \leftarrow B$
        if type(e'_j) = adj **then**
            $B_{adj} \leftarrow B_{adj} + (B - costs)$
        else
            $B_{nfeat} \leftarrow B_{nfeat} + (B - costs)$
        end if
    else
        ▷ Finish filling up the cache budget.
        Break
    end if
end for
return B_{adj}, B_{nfeat}, and x_i

If the last entry of E is partially cached, we can divide the last entry into two parts and treat them as two entries instead. Then the inequality above still holds. In conclusion, Algorithm 3 is always at least as better as the optimal solution at each iteration. Therefore by induction, Algorithm 3 will terminate with an optimal solution. □

Cache Constructor

In this section, we elaborate on the procedures for constructing both Adj-Cache and Nfeat-Cache. Since Algorithm 3 has already selected the specific entries to cache, our focus here is on detailing the layout and structural design of the two caches rather than the selection process itself.

Adj-Cache

Adj-Cache places frequently accessed portions of the adjacency matrix directly in GPU memory, thereby reducing CPU–GPU data transfers and speeding up Adj-Sampling. As in prior work [9, 11, 12], we employ a static caching strategy to avoid the significant overhead associated with dynamically tracking and swapping data between CPU and GPU. Designing Adj-Cache is non-trivial for two reasons: (1) adjacency lists have variable lengths, which may necessitate extra structures or logic to manage how cached entries are represented and queried, and (2) the access pattern of Adj-Sampling demands that cache lookups be extremely lightweight. Because adjacency-list accesses are interwoven with fine-grained CUDA kernel computations (see Sect. 2.2), it is essential that cache lookups do not hinder kernel execution. We tackle both issues by using a CSC-style layout for Adj-Cache, paired with graph reordering. This enables a design with no auxiliary lookup structures and a very cheap cache-hit check.

Concretely, given k adjacency lists selected for caching, we first reorder the graph so that these k lists appear at the beginning of the overall adjacency sequence. After reordering, $col\_index[0\colon k+1]$ (where $[0\colon k+1]$ denotes indices $0, 1, \ldots, k$) records the lengths of these lists, while $row\_index[0 : col\_index[k]]$ and $values[0 : col\_index[k]]$ contain the row indices and corresponding values for all cached lists. We then slice off these prefixes to obtain three cached CSC arrays: col_index_cached, row_index_cached, and $values\_cached$. If only part of the last entry is cached, we store the leading portion of that entry in row_index_cached and $values\_cached$, while col_index_cached remains $col\_index[0\colon k+1]$ because $col\_index[k]$ is indivisible. These three arrays constitute the Adj-Cache.

For illustration, suppose we want to cache the adjacency lists of node 0 and node 1 in the graph of Fig. 2.2. Since these lists already appear first, no reordering is required. We then slice the CSC arrays to obtain $col\_index\_cached = [0, 1, 3]$, $row\_index\_cached = [2, 2, 4]$, and $values\_cached = [1, 1, 1]$. Together, these arrays form Adj-Cache.

Both original and cached CSC arrays are provided to the sampling kernel. When the kernel requests the jth element from an original array, the lookup hits the cache whenever $j < L$, where L is the length of the cached array. On a hit, data is fetched directly from GPU memory. Otherwise, the kernel retrieves data from the original host-side CSC array via UVA.

This design is tailored to the properties of adjacency lists and their access pattern. First, it eliminates the need for a separate mapping structure to record offsets of variable-length cached entries, i.e., packing them contiguously via reordering and storing them in CSC form avoids additional lookup overhead. Second, because adjacency-list accesses occur alongside fine-grained computation (see Sect. 2.2), we make the cache lookup as minimal as a simple arithmetic comparison. To summarize, our approach yields an Adj-Cache that is both lightweight and highly efficient.

Nfeat-Cache

We adopt the same design for Nfeat-Cache as used in prior studies [9, 11]. Concretely, once the set of node features selected for caching is determined, we pack these features into a contiguous GPU tensor. Alongside this tensor, we maintain a lookup table that records the GPU-side address information for every node feature.

During Nfeat-Selecting, each requested node feature is first checked against this lookup table. If a query results in a cache hit, the feature is retrieved directly from

```
import DUCATI
# construct two caches
adj_cache, nfeat_cache = DUCATI.DualCacheAllocator(...)
# create and use adj sampler
sampler = DUCATI.NeighborSampler(fanouts, adj_cache)
in_node, out_node, batch_adj = sampler.sample(adj, seeds)
# create and use nfeat loader
loader = DUCATI.NfeatLoader(nfeat, nfeat_cache)
batch_nfeat = loader.load(node_ids)
```

Fig. 3.12 APIs of DUCATI for Adj-Sampling and Nfeat-Selecting

GPU memory using the stored address. When a cache miss occurs, the feature is instead pulled from host memory via the UVA mechanism. This setup ensures fast access for frequently used features while preserving correctness for uncached ones.

Trainer

Following prior systems [9, 13], DUCATI introduces only lightweight modifications to the DGL programming interface. Once Adj-Cache and Nfeat-Cache are built, DUCATI exposes two cache-related APIs, `sample` and `load`, as shown in Fig. 3.12. The `sample` API behaves exactly the same as in DGL, while the `load` API offers a simple and intuitive way for users to retrieve node features through the dual-cache mechanism.

Most parts of the training pipeline remain unchanged, except that DUCATI replaces the original Adj-Sampling and Nfeat-Selecting APIs. During these two stages, DUCATI first checks whether Adj-Cache or Nfeat-Cache hits. On a cache hit, the requested data is fetched directly from GPU memory; on a miss, the data is retrieved from host memory via the UVA mechanism. All forward computation, backward propagation, and weight updates of the Graph Neural Network are still handled by DGL without modification.

3.3.3 Experiments

We implemented DUCATI on top of DGL v0.8 and PyTorch v1.9, and released the source code on GitHub.[8] The full implementation contains 1.6k lines of Python code and 422 lines of CUDA/C++ code. All experiments were conducted on a single machine equipped with one NVIDIA GeForce RTX 2080 Ti GPU (11GB HBM), PCIe 3.0$\times$16, 256GB DDR4 memory, and an Intel(R) Xeon(R) Gold 6240 CPU running at 2.60GHz.

[8] https://github.com/initzhang/DUCATI_SIGMOD.git.

Table 3.10 Dataset statistics

Dataset	Nodes	Edges	Feature	Size(adj)	Size(nfeat)
PA	111M	1.6B	128/256	12.9GB	54/108GB
UK	77.7M	3.0B	128/256	11.3GB	33/75GB
TW	41.7M	1.5B	128/256	22.7GB	20/40GB
UU	134M	5.5B	128/256	42.0GB	65/130GB
PR	2.4M	124M	100	0.94GB	0.91GB
RD	232K	114M	602	0.86GB	0.13GB

The node features are stored as float32 type. The adjacency matrices are stored as int64 type in the CSC format

Experimental Setup

Our work aims to enhance the efficiency and scalability of GNN training on massive graphs. Consequently, we evaluate DUCATI under the most demanding mini-batch training scenario when both the adjacency matrix and node features exceed the capacity of a single GPU's memory. Prior studies [11, 24] consider a more limited environment in which the adjacency matrix is assumed to fit entirely within GPU memory. In contrast, our experiments impose no such requirement, providing a more realistic and comprehensive assessment of how different GNN systems scale in truly large-graph settings.

Datasets. Our evaluation uses four billion-scale datasets for assessing mini-batch training systems and two smaller datasets for full-batch systems, as summarized in Table 3.10. Because full-batch approaches scale poorly and lead to GPU OOM on billion-scale graphs (as discussed in Sect. 3.3.3), we limit the comparison with full-batch systems to the two smaller datasets. For large-scale experiments, we adopt two web graphs [25]: UK-2006-05 (UK) and UK-Union (UU), a social graph [26]: Twitter (TW), and a citation graph from the Open Graph Benchmark (OGB) [2]: Ogbn-Papers100M (PA). For smaller datasets, we use the co-purchasing network Ogbn-Products (PR) from OGB and the Reddit (RD) graph [8]. Following prior work [9, 11], we generate random node features for UK, UU, and TW since they contain none originally. To evaluate scalability and adaptability, we generate two common feature dimensions (128 and 256) per dataset. OGB provides an official training split for PA; for datasets lacking one, we sample 1% of the vertices as training data. Since nodes in the PA training set have higher average degree, we sample the training sets of the other datasets proportionally to their normalized degrees.

GNNs and Hyperparameters of Mini-batch Training. In line with prior works [9, 13], we benchmark DUCATI and baselines using DGL's three-layer GraphSAGE implementation, setting the hidden dimension to 16 as in [9]. We follow [11] and use a batch size of 8000 for every system. Two hyperparameters strongly affect mini-batch workload: the neighborhood fanouts for Adj-Sampling, and the node feature dimension for Nfeat-Selecting. Some practitioners [27] prefer smaller fanouts such as (2,2,2) for faster runs, while others [8] adopt larger ones (15,10,5) to improve accuracy. Different datasets also imply different feature dimensions. We summarize their effects as follows:

- Workload of Nfeat-Selecting: reducing fanouts or decreasing feature dimension lowers the number and size of features in each mini-batch, thereby reducing workload.
- Workload of Adj-Sampling: while smaller fanouts reduce the number of sampled nodes, the overall cost decreases more slowly than in Nfeat-Selecting, because the random permutations required for sampling are fanout-independent. Feature dimension does not affect Adj-Sampling.

To evaluate adaptability, we test four settings that yield diverse Adj-Sampling and Nfeat-Selecting workloads:

- fanouts = (2,2,2), dimension = 128;
- fanouts = (2,2,2), dimension = 256;
- fanouts = (15,10,5), dimension = 128;
- fanouts = (15,10,5), dimension = 256.

Baselines. We group the baselines into five categories:

- **Base System:** PyG [5], which performs both Adj-Sampling and Nfeat-Selecting on the CPU.
- **System with UVA:** DGL [17]. The latest v0.8 release supports UVA but incorporates no caching.
- **System with Single-Cache:** PaGraph [9] and GNNLab [11]. PaGraph relies on CPU-based sampling, which is significantly slower than UVA-based GPU sampling [12]. GNNLab cannot handle any dataset in Table 3.10 because it requires the adjacency matrix to fit entirely in GPU memory.
- **System with UVA and Single-Cache:** Quiver [12] and a synthetic system SOTA. Quiver combines UVA-based Adj-Sampling with Nfeat-Cache. SOTA is implemented using the same code base as DUCATI and incorporates UVA-based Adj-Sampling, UVA-based Nfeat-Selecting, and the GNNLab-style Nfeat-Cache, which is more effective than other cache implementations [11].
- **System with UVA and Dual-Cache:** the proposed DUCATI.

We include SOTA for two reasons: (1) since it shares the same infrastructure as DUCATI, comparing the two isolates the benefit of our Dual-Cache over a strong Single-Cache design; and (2) SOTA outperforms prior systems (validated in Sect. 3.3.3), enabling us to focus comparisons on only DUCATI, SOTA, and DGL. Following [9, 11], we determine the total cache size dynamically by monitoring peak GPU memory during several initial batches and allocating the remaining GPU memory to caching. DUCATI and SOTA always receive identical cache budgets. We report the iteration time computed over 1000 batches for each configuration, i.e., the average wall-clock time per mini-batch.

Table 3.11 Iteration time of all methods (unit: ms)

Dataset	Setting	DGL	*SOTA*	DUCATI
PA	2,2,2 + 128	16.05 (1.51x)	13.40 (1.26x)	10.65
PA	2,2,2 + 256	18.48 (1.82x)	12.75 (1.26x)	10.13
PA	15,10,5 + 128	49.24 (2.67x)	28.39 (1.54x)	18.45
PA	15,10,5 + 256	71.79 (3.33x)	31.16 (1.44x)	21.59
UK	2,2,2 + 128	18.54 (1.60x)	15.58 (1.34x)	11.62
UK	2,2,2 + 256	23.63 (1.65x)	18.79 (1.31x)	14.35
UK	15,10,5 + 128	61.41 (2.22x)	35.72 (1.29x)	27.64
UK	15,10,5 + 256	99.61 (2.15x)	61.56 (1.33x)	46.37
TW	2,2,2 + 128	17.81 (1.29x)	15.63 (1.13x)	13.85
TW	2,2,2 + 256	19.86 (1.50x)	15.44 (1.17x)	13.22
TW	15,10,5 + 128	71.74 (2.95x)	33.06 (1.36x)	24.30
TW	15,10,5 + 256	119.89 (3.00x)	53.43 (1.33x)	40.03
UU	2,2,2 + 128	18.47 (1.55x)	15.95 (1.34x)	11.90
UU	2,2,2 + 256	23.56 (1.61x)	19.87 (1.36x)	14.63
UU	15,10,5 + 128	62.83 (2.15x)	39.95 (1.37x)	29.21
UU	15,10,5 + 256	100.00 (2.04x)	66.95 (1.36x)	49.11
	Avg. Speedup	2.07x	1.32x	1.00x

Setting represents the neighbor fanouts + node features dimension

Evaluation of Overall Training Time

The end-to-end per-iteration training times of DUCATI and all baselines are summarized in Table 3.11. Overall, DUCATI delivers an average acceleration of 1.32 times compared with SOTA and 2.07 times compared with DGL. The speedup can be observed under all settings, which verifies the generalization ability of DUCATI.

(1) Comparison with DGL. The results highlight the advantage of retaining frequently accessed graph data directly in GPU memory. When both the fanout values and the feature dimensions are small, the amount of graph data per mini-batch is limited, so the overhead of repeatedly transferring data via UVA is less pronounced, resulting in modest gains for SOTA and DUCATI. As the fanouts and feature dimensions increase, however, DGL suffers significantly due to intensive data movement, while the caching strategies adopted by SOTA and DUCATI mitigate this bottleneck and yield notably larger speedups.
(2) Comparison with SOTA. The results also reveal the benefit of the Dual-Cache approach relative to the Single-Cache design. Even though DUCATI and SOTA use the same total GPU memory for caching, DUCATI achieves superior throughput, indicating that it exploits the available memory more effectively. This improvement stems from the fact that Nfeat-Cache alone does not fully utilize the remaining GPU memory, whereas DUCATI's Adj-Cache leverages this leftover memory to accelerate the computationally heavy Adj-Sampling step.

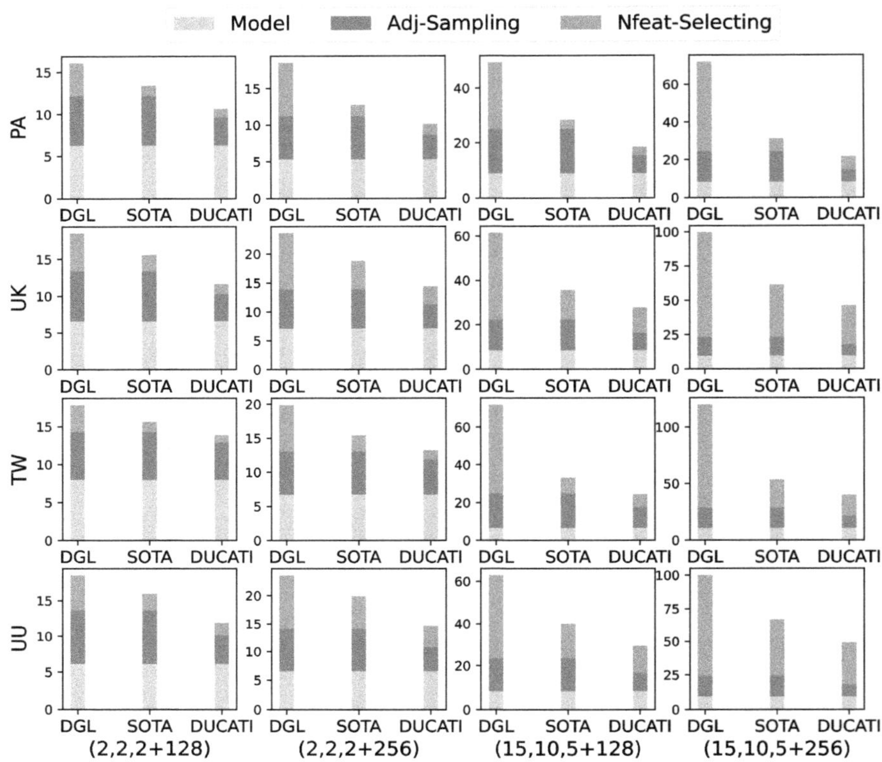

Fig. 3.13 Breakdown of iteration time (unit of Y-axis: ms)

(3) A cross-setting comparison shows that the relative advantage of DUCATI over both SOTA and DGL generally becomes more significant as the input size (fanouts and feature dimensions) increases. Put differently, as the workload grows, DUCATI's iteration time increases more slowly than that of the baselines. This trend demonstrates the strong scalability of our system.

Training Time Breakdown

We further decompose the iteration time in Fig. 3.13. Each iteration is made up of three components: the execution time of Adj-Sampling, Nfeat-Selecting, and the Model stage (forward pass, backward propagation, and parameter updates). Since all three systems generate identical mini-batches under the same configuration, the Model cost remains unchanged across methods. Thus, the performance differences originate entirely from the Adj-Sampling and Nfeat-Selecting stages.

(1) Adj-Sampling. DUCATI demonstrates a substantial advantage over both SOTA and DGL due to the introduction of the Adj-Cache. By storing frequently touched adjacency entries directly in GPU memory, DUCATI significantly cuts down data movement within Adj-Sampling, resulting in markedly lower sampling time.
(2) Nfeat-Selecting. DUCATI achieves nearly the same efficiency as SOTA in this stage. Although DUCATI dedicates part of the cache capacity to Adj-Cache and therefore maintains a slightly smaller Nfeat-Cache, the reduction has minimal performance impact because of the previously observed diminishing returns of Nfeat-Cache (Sect. 3.3.1). The Dual-Cache Allocator intentionally divides the cache space so that both caches store only the most beneficial entries.

Evaluation of Dual-Cache Allocator

To examine the behavior of the Dual-Cache Allocator, we perform a case study by enumerating all possible cache partition plans under a given total budget and measuring the performance of each plan. We conduct this exhaustive evaluation on four datasets and present the results as heatmaps in Fig. 3.14.[9] Detailed cache-partition choices made by the allocator are provided in the GitHub repository due to space constraints.

(1) The allocator's selected plans consistently lie within the optimal regions of the heatmaps, confirming its effectiveness. Because the node feature tensor typically occupies far more space than the adjacency structure, assigning a larger fraction of the budget to Nfeat-Cache is usually desirable. Consequently, the green pentagrams marking the allocator's choices appear toward the left side of each heatmap.
(2) The conventional Single-Cache baseline always corresponds to the leftmost rectangles, allocating the entire budget to the Nfeat-Cache. As shown in the figure, such plans are consistently suboptimal. In contrast, the Dual-Cache approach reallocates a portion of this budget to Adj-Cache, which leads to noticeable reductions in the total training time.
(3) We also observe that multiple allocation plans may produce nearly identical performance. For instance, on Ogbn-Papers100M, several dark rectangles appear in close proximity, indicating that their performance differs by less than 0.1 ms. This suggests that the allocation landscape contains a smooth region around the local optimum, making it easy for the Dual-Cache Allocator to locate high-quality solutions.

Finally, the allocator remains effective even under extremely constrained budgets. When we manually restrict the total cache capacity to only 0.2,GB or 0.4,GB,

[9] We use fanouts (15,10,5) and a node feature dimension of 128 for this case study; other configurations show similar trends. A 0.2, GB step size is used for the grid search.

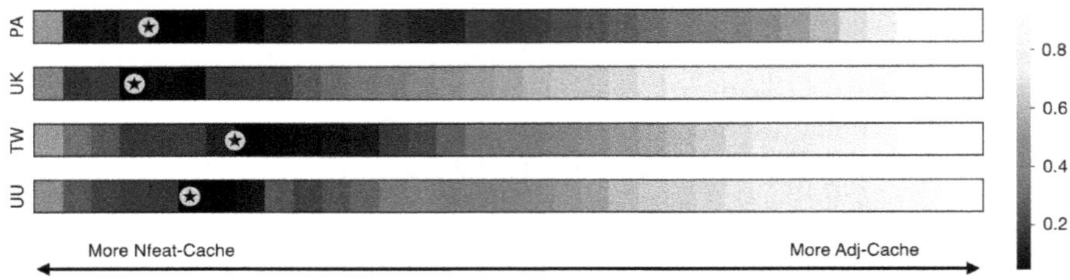

Fig. 3.14 The heatmap that visualizes the normalized execution time of all possible allocation plans. Here, each rectangle represents one allocation plan. The position of the rectangle indicates the size of Adj-Cache. From left to right, the size of Adj-Cache increases from zero to the total available cache budget with a fixed step size, and the size of Nfeat-Cache decreases accordingly. The hollow green pentagram denotes the allocation plan given by our Dual-Cache Allocator. The darker color means shorter time

DUCATI still outperforms SOTA, achieving iteration times of 33.39/26.09 ms versus SOTA's 40.90/36.72 ms.[10]

Evaluation of Preprocessing Time

In this section, we examine the overhead introduced by the Input Inspector, Dual-Cache Allocator, and Cache Constructor. We treat the sum of their execution times as the preprocessing cost and compare it against the per-epoch training time in Table 3.12. Our position is that this preprocessing cost can be effectively amortized over the course of training.

Prior studies [11, 28, 29] show that achieving satisfactory accuracy on extremely large graphs typically involves training for hundreds or even thousands of epochs. Across all datasets, DUCATI requires fewer than 100 epochs' worth of time for preprocessing. Moreover, community practice indicates that users frequently adopt smaller batch sizes and larger hidden-layer widths for GNN models on public leaderboards.[11] These choices significantly increase the per-epoch training time while having negligible impact on the preprocessing stage. For instance, on PA, when using (batch size, hidden layer size) of (4000,512) or (1000,1024), the epoch time becomes 9 s or 23 s, respectively, whereas the preprocessing cost is only 101 s or 103 s. Consequently, the preprocessing cost of DUCATI is minor in comparison with the overall training time.

[10] PA dataset, (15,10,5 + 128) configuration.

[11] https://ogb.stanford.edu/docs/leader_nodeprop/#ogbn-papers100M.

Table 3.12 Comparison of the preprocessing and the epoch time of DUCATI on four datasets under (15,10,5 + 128) setting

Dataset	Prep. Time (s)	Epoch time (s)
PA	105.2 ± 2.1	2.77 ± 0.03
UK	81.0 ± 1.5	2.68 ± 0.02
TW	78.3 ± 1.7	1.24 ± 0.02
UU	153.2 ± 3.4	4.88 ± 0.05

Table 3.13 The performance comparison of DUCATI, MariusGNN, and Quiver for node classification on PA using a three-layer GraphSAGE model

System	Accuracy	Epoch time (s)
DUCATI	0.670 ± 0.003	28.9 ± 0.5
SOTA	0.671 ± 0.001	34.8 ± 0.3
Quiver	0.665 ± 0.002	58.0 ± 3.1
MariusGNN	0.651 ± 0.002	92.3 ± 12.4

Comparison with Mini-Batch Systems

In this section, we benchmark DUCATI against two recent mini-batch GNN training systems, MariusGNN [22] and Quiver [12], in terms of both accuracy and training throughput. We also include the performance of SOTA to confirm that it indeed represents a strong baseline. Following prior work [9, 11, 12], we assess system efficiency using the standard node-classification task on PA. All four systems are evaluated using a single 2080Ti GPU (11GB). Following MariusGNN, we augment PA with reverse edges, which nearly doubles the number of edges. We adopt the GraphSAGE model configuration provided in the publicly released MariusGNN artifact.[12] As in MariusGNN, we set the batch size to 1000 and the hidden dimension to 256, and we report both peak validation accuracy and average epoch time. The fanouts for all systems are fixed at (10,10,10). Each metric is reported as the mean and confidence interval over three runs, each consisting of twenty epochs.

As summarized in Table 3.13, DUCATI achieves the shortest training time among all four systems, while MariusGNN produces noticeably lower accuracy than DUCATI, SOTA, and Quiver. These observations can be explained by several fundamental architectural differences.

First, MariusGNN is designed as a disk-oriented system that emphasizes CPU-memory caching and on-disk data layout. In contrast, DUCATI is built as a memory-centric framework that explicitly utilizes GPU-resident caches and GPU-friendly data structures. Even when CPU and GPU memory are ample, MariusGNN keeps all input data in host memory and does not take advantage of GPU caching. Consequently, the high CPU–GPU data-transfer overhead significantly slows down MariusGNN in

[12] https://github.com/marius-team/marius/tree/eurosys_2023_artifact.

Table 3.14 Performance comparison of DUCATI, Sancus, and DGL(SF) for node classification on PR and RD using a three-layer GCN model

System	Converg. Time (s)		Accuracy	
	PR	RD	PR	RD
DUCATI	173.7 ± 2.3	95.1 ± 1.2	0.913 ± 0.000	0.931 ± 0.001
Sancus	691.6 ± 1.4	142.6 ± 0.2	0.912 ± 0.000	0.935 ± 0.000
DGL(SF)	OOM	30.0 ± 0.9	OOM	0.938 ± 0.001

comparison with DUCATI. Furthermore, MariusGNN alters the standard neighbor-sampling routine to reduce sampling overhead. However, this modification weakens sampling randomness, which contributes to its reduced accuracy.

Second, MariusGNN's disk-based approach offers a major benefit. It can scale beyond the limits of CPU memory. For instance, on the largest AWS P3 node (488GB host memory, 16TB local storage), DUCATI cannot process the trillion-edge, 8.5TB Facebook15 graph [30], whereas MariusGNN can [22]. In scenarios where a slight accuracy reduction is acceptable, the modified sampling strategy of MariusGNN may further improve efficiency. By contrast, SOTA and Quiver, though memory-based and capable of exploiting GPU caching, do not leverage locality in the adjacency structure and are still subject to the diminishing returns of Nfeat-Cache discussed in Sect. 3. As a result, they underutilize the available GPU memory and therefore lag behind DUCATI in terms of speed. Regarding accuracy, SOTA, Quiver, and DUCATI all employ the unmodified neighbor-sampling algorithm and thus obtain comparable performance.

Comparison with Full-Batch Systems

In this section, we benchmark DUCATI against two state-of-the-art full-batch GNN training frameworks, Sancus [10] and DGL(SF) [17], in terms of accuracy and training efficiency. Sancus performs full-batch training across multiple GPUs, whereas DGL(SF) adopts a full-batch pipeline on a single GPU. Because DUCATI is a mini-batch system but Sancus and DGL(SF) are full-batch systems, we report convergence time rather than per-epoch time to ensure fairness. DUCATI, DGL(SF), and Sancus use one, one, and eight 2080Ti (11GB) GPUs, respectively. We employ the GCN model used in Sancus's released implementation.[13] Following Sancus, we set the training epochs to 500/300 and hidden dimension to 128/16 for PR/RD. DGL(SF) adopts the same configuration. For DUCATI, we match the hidden dimension, use fanouts (15,15,15), set the batch size to 1000, and train for 30 epochs. Each metric is reported as the mean with confidence intervals from three independent runs. The results are shown in Table 3.14.

[13] https://github.com/chenzhao/light-dist-gnn.

First, both Sancus and DGL(SF) fail with GPU OOM on the four billion-scale graphs when using eight and one 2080Ti (11GB), respectively. In contrast, these datasets can be trained without issue by DUCATI, MariusGNN, or Quiver using just one 2080Ti. This confirms that, under limited GPU resources, full-batch systems scale significantly worse than mini-batch approaches. This is reasonable since full-batch pipelines must keep the entire graph and the hidden representations in GPU memory, dramatically increasing memory requirements relative to mini-batch training.

Second, while DGL(SF) cannot scale beyond a single GPU, Sancus distributes training across multiple GPUs to improve scalability. However, this comes at the cost of reduced accuracy. The inter-GPU communication is sufficiently expensive that Sancus introduces staleness during training to maintain throughput, which leads to noticeable accuracy degradation as shown in Table 3.14. Moreover, Sancus must still hold the full dataset and hidden activations across the GPUs, which continues to limit the scalability to billion-scale graphs like PA when only modest GPU resources are available.

Third, DUCATI and other mini-batch systems intentionally trade accuracy for scalability by using neighbor sampling. On one hand, sampling inevitably removes part of the neighborhood information, explaining DUCATI's lower accuracy relative to Sancus in Table 3.14. On the other hand, sampling drastically reduces GPU memory usage, because only the sampled mini-batch must reside in GPU memory during training.

3.4 Conclusion

In this chapter, we propose the feature-oriented sampling framework designed to enhance the scalability and efficiency of GNN learning on massive graphs. We point out three fundamental issues in conventional topology-oriented sampling approaches that hinder scalability, and we show that the FOS sampler effectively resolves these issues. Applying FOS to three representative GNN architectures (GCN, GraphSAGE, and GAT) and evaluating it on four real-world large datasets (Reddit, Yelp, Ogbn-Products, and Amazon), our experiments demonstrate that FOS attains accuracy comparable to traditional TOS-based methods while accelerating training by a factor of $2.2 \sim 7.9$. We further introduce DUCATI, a Dual-Cache training system that further boosts the scalability and efficiency of mini-batch GNN training on billion-scale graphs. We identify three key shortcomings in existing systems: insufficient exploitation of adjacency locality, underutilization of available GPU memory, and limited adaptability across workloads. DUCATI tackles these issues through the introduction of Adj-Cache and a workload-aware Dual-Cache Allocator. Comprehensive evaluations on four large datasets (Ogbn-Papers100M, UK-2006-05, Twitter, and UK-Union) show that DUCATI reduces training time by up to 3.33× (with an average of 2.07×) over DGL, and by up to 1.54× (with an average of 1.32×) compared with state-of-the-art Single-Cache approaches. In addition, we benchmark

DUCATI against four leading GNN training systems based on accuracy-time tradeoffs. The results indicate that when the overall training memory footprint exceeds GPU capacity but remains within CPU limits, DUCATI provides the most effective solution for mini-batch GNN training.

References

1. Thomas N. Kipf and Max Welling. 2017. Semi-Supervised Classification with Graph Convolutional Networks. In *ICLR*. https://openreview.net/forum?id=SJU4ayYgl
2. Weihua Hu, Matthias Fey, Marinka Zitnik, Yuxiao Dong, Hongyu Ren, Bowen Liu, Michele Catasta, and Jure Leskovec. 2020. Open graph benchmark: Datasets for machine learning on graphs. *NeurIPS* 33 (2020), 22118–22133.
3. Hanqing Zeng, Hongkuan Zhou, Ajitesh Srivastava, Rajgopal Kannan, Viktor K. Prasanna. 2020. GraphSAINT: Graph sampling based inductive learning method. In *ICLR*.
4. Difan Zou, Ziniu Hu, Yewen Wang, Song Jiang, Yizhou Sun, Quanquan Gu. 2019. Layer-dependent importance sampling for training deep and large graph convolutional networks. *NeurIPS*. 11247-11256.
5. Matthias Fey, Jan Eric Lenssen. 2019. Fast graph representation learning with PyTorch Geometric. In *ICLR Workshop*.
6. Rex Ying, Ruining He, Kaifeng Chen, Pong Eksombatchai, William L. Hamilton, Jure Leskovec. 2018. Graph convolutional neural networks for web-scale recommender systems. In *KDD*. 257–266.
7. Wei-Lin Chiang, Xuanqing Liu, Si Si, Yang Li, Samy Bengio, Cho-Jui Hsieh. 2019. Cluster-GCN: An efficient algorithm for training deep and large graph convolutional networks. In *KDD*. 257–266.
8. Will Hamilton, Zhitao Ying, and Jure Leskovec. 2017. Inductive representation learning on large graphs. *Advances in neural information processing systems* 30 (2017).
9. Zhiqi Lin, Cheng Li, Youshan Miao, Yunxin Liu, and Yinlong Xu. 2020. Pagraph: Scaling gnn training on large graphs via computation-aware caching. In *SoCC*. 401–415.
10. Jingshu Peng, Zhao Chen, Yingxia Shao, Yanyan Shen, Lei Chen, and Jiannong Cao. 2022. SANCUS: Staleness-Aware Communication-Avoiding Full-Graph Decentralized Training in Large-Scale Graph Neural Networks. *VLDB* 15, 9 (2022), 1937–1950. https://doi.org/10.14778/3538598.3538614
11. Jianbang Yang, Dahai Tang, Xiaoniu Song, Lei Wang, Qiang Yin, Rong Chen, Wenyuan Yu, and Jingren Zhou. 2022. GNNLab: a factored system for sample-based GNN training over GPUs. In *EuroSys*. 417–434.
12. Zeyuan Tan, Xiulong Yuan, Congjie He, Man-Kit Sit, Guo Li, Xiaoze Liu, Baole Ai, Kai Zeng, Peter R. Pietzuch, Luo Mai. 2023. Quiver: Supporting GPUs for Low-Latency, High-Throughput GNN Serving with Workload Awareness. *arXiv preprint* arXiv:2305.10863.
13. Seungwon Min, Kun Wu, Sitao Huang, Mert Hidayetoglu, Jinjun Xiong, Eiman Ebrahimi, Deming Chen, Wen-mei W. Hwu. 2021. Large graph convolutional network training with gpu-oriented data communication architecture. *PVLDB*. 14(11): 2087-2100.
14. Jialin Dong, Da Zheng, Lin F. Yang, George Karypis. 2021. Global Neighbor Sampling for Mixed CPU-GPU Training on Giant Graphs. In *KDD*. 289-299.
15. Jiezhong Qiu, Jian Tang, Hao Ma, Yuxiao Dong, Kuansan Wang, Jie Tang. 2018. Deepinf: Social influence prediction with deep learning. In *KDD*. 2110–2119.
16. Cheng Wan, Youjie Li, Cameron R. Wolfe, Anastasios Kyrillidis, Nam Sung Kim, Yingyan Lin. 2022. PipeGCN: Efficient full-graph training of graph convolutional networks with pipelined feature communication. In *ICLR*.

17. Minjie Wang, Da Zheng, Zihao Ye, Quan Gan, Mufei Li, Xiang Song, Jinjing Zhou, Chao Ma, Lingfan Yu, Yu Gai, et al. 2019. Deep graph library: A graph-centric, highly-performant package for graph neural networks. *arXiv preprint* arXiv:1909.01315 (2019).
18. Adam Paszke, Sam Gross, Soumith Chintala, Gregory Chanan, Edward Yang, Zachary DeVito, Zeming Lin, Alban Desmaison, Luca Antiga, Adam Lerer. 2017. Automatic differentiation in PyTorch.
19. Jie Chen, Tengfei Ma, Cao Xiao. 2018. FastGCN: Fast Learning with Graph Convolutional Networks via Importance Sampling. In *ICLR*.
20. Carl Pearson, Abdul Dakkak, Sarah Hashash, Cheng Li, I-Hsin Chung, Jinjun Xiong, Wen-Mei Hwu. 2019. Evaluating characteristics of CUDA communication primitives on high-bandwidth interconnects. In *ICPE*. 209-218.
21. Diederik P. Kingma, Jimmy Ba. 2015. Adam: A method for stochastic optimization. In *ICLR*.
22. Roger Waleffe, Jason Mohoney, Theodoros Rekatsinas, Shivaram Venkataraman. 2023. MariusGNN: Resource-Efficient Out-of-Core Training of Graph Neural Networks. In *EuroSys*.
23. Ang Li, Shuaiwen Leon Song, Jieyang Chen, Jiajia Li, Xu Liu, Nathan R. Tallent, Kevin J. Barker. 2020. Evaluating modern gpu interconnect: Pcie, nvlink, nv-sli, nvswitch and gpudirect. *IEEE Trans. Parallel Distrib. Syst.*31(1): 94-110.
24. Abhinav Jangda, Sandeep Polisetty, Arjun Guha, Marco Serafini. 2021. Accelerating graph sampling for graph machine learning using GPUs. In *EuroSys*. 311-326.
25. Paolo Boldi, Sebastiano Vigna. 2004. The webgraph framework I: compression techniques. In *WWW*. 595-602.
26. Haewoon Kwak, Changhyun Lee, Hosung Park, Sue B. Moon. 2010. What is Twitter, a social network or a news media? In *WWW*. 591-600.
27. Jianfei Chen, Jun Zhu, Le Song. 2017. Stochastic training of graph convolutional networks with variance reduction. *ICML*. 941-949.
28. Swapnil Gandhi and Anand Padmanabha Iyer. 2021. P3: Distributed deep graph learning at scale. In *OSDI*. 551–568.
29. Yuke Wang, Boyuan Feng, Gushu Li, Shuangchen Li, Lei Deng, Yuan Xie, and Yufei Ding. 2021. GNNAdvisor: An adaptive and efficient runtime system for GNN acceleration on GPUs. In *OSDI*. 515–531.
30. Avery Ching, Sergey Edunov, Maja Kabiljo, Dionysios Logothetis, Sambavi Muthukrishnan. 2015 One trillion edges: Graph processing at facebook-scale. *Proc. VLDB Endow*. 8(12): 1804-1815.

Chapter 4
Accelerating Computation for Scalable Graph Neural Network Training

Jingzhi Fang and Zhiyuan Li

Abstract This chapter addresses computational acceleration techniques for scalable GNN training. At the model level, prevalent frameworks incur significant overhead from batch preparation and CPU-GPU data transfer in mini-batch training. Furthermore, their constrained device utilization patterns restrict opportunities for pipeline parallelism. To overcome these limitations, we propose DAHA, a GNN training framework featuring data-aware and hardware-aware execution planning that optimizes end-to-end training workflows. Extensive experiments confirm that DAHA delivers consistent and substantial speedups while generalizing across diverse message-passing GNN architectures. At the operator level, prior approaches for accelerating sparse operators often decompose sparse tensors and search for hybrid format combinations tailored to specific sparsity patterns. However, these methods face a fundamental trade-off between search space size and exploration time, which can limit achievable efficiency. To resolve this, we propose STile, a framework that expands the optimization space both in breadth through flexible sparse tensor transformations and in depth via multi-level decomposition. Empirical evaluations of SpMM and SDDMM operators across varied sparsity patterns demonstrate the effectiveness of STile.

4.1 Background

The demonstrated efficacy of GNNs [19] in domains such as network embedding [21] and recommendation systems [30] has led to their widespread adoption. Consequently, a central and persistent research challenge lies in developing efficient training methodologies. To address this, mini-batch sampling techniques [4–7, 12, 15, 32, 33] have emerged as a popular paradigm. The unique computational demands of graph-structured data necessitate specialized frameworks like PyG [8] and DGL [26],

---

J. Fang (✉) · Z. Li
The Hong Kong University of Science and Technology, Hong Kong SAR, China
e-mail: jfangak@connect.ust.hk

Z. Li
e-mail: zlicw@connect.ust.hk

Y. Shen and L. Chen (eds.), *Graph Neural Network Training*, Machine Learning: Foundations, Methodologies, and Applications,
https://doi.org/10.1007/978-981-95-5795-0_4

which are designed for GPU acceleration, as conventional deep learning systems cannot handle the irregular data access patterns inherent to GNN operations.

The prevailing training architecture for GNNs delineates roles between (1) the CPU, responsible for batch sampling and feature slicing and (2) the GPU, which performs the core model computations. This separation creates a critical pipeline dependency. Slow batch preparation on the CPU can idle the GPU [35], while the substantial volume of data transferred, driven by neighbor explosion and large feature dimensions, imposes a major communication cost [10, 33]. Empirical studies thus pinpoint batch preparation and CPU-GPU data transfer as the twin bottlenecks in modern GNN training systems [10, 16, 20, 24, 27]. Their sequential position in the pipeline means they directly throttle downstream GPU operations, leading to suboptimal hardware utilization. While recent studies have started to investigate alternative device utilization patterns, the explored design space remains constrained. Notably, they overlook potential optimizations in pipeline parallelism, particularly those achievable through fine-grained operator scheduling.

On the operator level, sparse operators, i.e., operators that take sparse tensors as input, are a kind of important operators in GNN to be optimized. For example, the Sparse Matrix Multiplication (SpMM) operator computes the multiplication result of a sparse matrix and a dense matrix. Compared with dense matrix multiplication, SpMM can skip the multiply-add operations related to the zeros of the sparse matrix. Another example is Sampled Dense-Dense Matrix Multiplication (SDDMM), which computes dense matrix multiplication but only samples a part of the result as output. In GNN, the sparse adjacent matrices [51, 67, 69] bring a variety of sparsity patterns in the tensors.

A number of sparse formats have been developed to cope with the diverse sparse patterns in the sparse operators. These formats have advantages and disadvantages when dealing with different sparsity patterns. It is well-recognized that no single format can suit all sparse tensors. Moreover, if a sparse tensor has different sparsity patterns in different regions, we may need to combine different formats together to achieve optimal performance. To deal with the diverse sparsity patterns of different sparse tensors and of different parts of a sparse tensor, people try to use hybrid formats, i.e., taking the sparse operator into consideration, they decompose a sparse tensor into several sparse sub-tensors and select the best sparse format for each of them [48, 55, 65, 66, 76, 78], then they schedule the computation based on the hybrid format. However, the existing works often make a trade-off between search space and search time. As a result, their search spaces are limited in some cases, limiting the optimized operator efficiency they can achieve.

In this chapter, we first propose a GNN training framework, DAHA, with data- and hardware-aware execution planning to accelerate end-to-end GNN training in Sect. 4.2. Then we propose a framework STile in Sect. 4.3, which extends the search space in its breadth (by doing flexible sparse tensor transformations) and depth (by enabling multi-level decomposition) to achieve higher operator running efficiency.

4.2 Data and Hardware-Aware Execution Planning

4.2.1 Motivation

As mentioned in Sect. 4.1, the CPU-sample-GPU-train execution in most of the GNN frameworks can result in low training efficiency. To mitigate these costs, recent frameworks such as DistDGLv2 [36] and SALIENT [16] employ pipelining to overlap batch preparation, transfer, and computation. Nonetheless, their adherence to the standard CPU-for-sampling/GPU-for-training model means performance remains fundamentally bounded by the slowest sequential operation in this chain. While newer systems explore different hardware roles, ByteGNN [35] for CPU efficiency, GNNLab [31] for GPU optimization, and a hybrid aggregation method by Kim et al. [18], they either underutilize one device, face CPU bottlenecks, or maintain coarse-grained scheduling. Critically, most prior work schedules entire batches monolithically, missing opportunities to exploit fine-grained operator-level pipeline parallelism.

We identify a critical research gap: the failure to exhaustively explore the combinatorial space of device utilization patterns at a fine-grained, operator level. This oversight misses significant opportunities to solve the underlying optimization problem of discovering the globally optimal execution plan. To address this, we propose to fill the gap by expanding the search space for efficiency optimization. Our approach enables a fully hybrid CPU-GPU paradigm, where each fine-grained operation within a batch can be dynamically assigned to any feasible device. The objective is to algorithmically discover the optimal execution plan, a solution conditioned on both the specific input data and the available hardware, following principles established in recent adaptive systems research [29].

Problem Statement. We focus on developing novel execution planning strategies for message-passing GNN training that are explicitly aware of the graph data's properties and the hardware's computational profile. Let G be an input graph, $\mathcal{A}$ be a GNN model, $\mathcal{B} = \mathcal{B}_1, ..., \mathcal{B}_n$ be a sequence of mini-batches (jobs), and $M = M_1, ..., M_m$ be a set of heterogeneous processing units (e.g., CPUs, GPUs). Our aim is to find an assignment and schedule (an execution plan) that maps the computational operations required to train $\mathcal{A}$ on $\mathcal{B}$ onto the machines in M, such that the makespan is minimized (Fig. 4.1).

4.2.2 A Data and Hardware-Aware Cost Model

DAHA employs a data- and hardware-aware cost model that predicts both neural computation time on CPU/GPU and data transfer time between them. The model focuses on two core operations: dense matrix multiplication (mm) and sparse matrix-matrix multiplication (spmm). By representing a GNN layer as a Message Flow Graph (MFG), the model key factors are the number of unique nodes (n_i) and edges

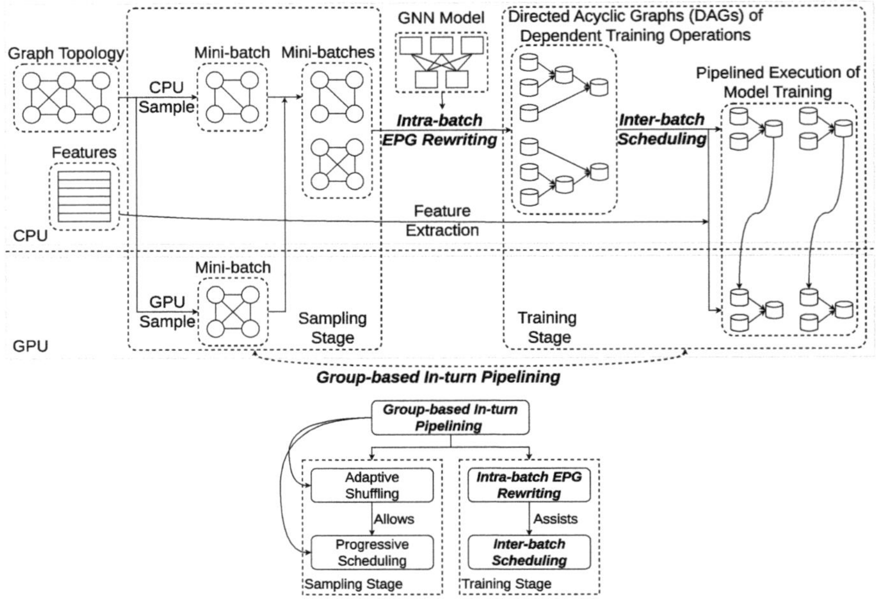

Fig. 4.1 DAHA architecture incorporates three new modules: (1) group-based in-turn pipelining, (2) intra-batch EPG (execution plan graph) rewriting, and (3) inter-batch scheduling. The pipelining module establishes the fundamental workflow by alternating sampling and training. Next, EPG rewriting prepares the search space, enabling the scheduling module to execute optimizations that enhance training speed

(m_i) in each MFG_i. The computation cost is modeled as a cubic regression on these variables (Eq. 4.1), where the regression parameters ($\{\beta_j\}$) encode both model architecture (e.g., hidden dimensions) and hardware performance. This formulation enables efficient, adaptive cost estimation tailored to specific data and hardware configurations. By fitting a dedicated set of parameters, we can subsequently estimate the computational cost for any individual operation within the pipeline.

$$\begin{aligned}\text{compCost}(\text{MFG}_i) &= \beta_0 + \beta_1 n_i + \beta_2 m_i + \beta_{11} {n_i}^2 + \beta_{22} {m_i}^2 \\ &+ \beta_{12} n_i m_i + \beta_{111} {n_i}^3 + \beta_{222} {m_i}^3 + \beta_{112} {n_i}^2 m_i + \beta_{122} n_i {m_i}^2.\end{aligned} \tag{4.1}$$

This per-MFG cost model generalizes naturally to the batch level. Given a batch $\mathcal{B}_j$ with MFGs $\{\text{MFG}_i\}$,

$$\text{compCost}(\mathcal{B}_j) = \sum_{\text{MFG}_i \in \mathcal{B}_j} \text{compCost}(\text{MFG}_i). \tag{4.2}$$

DAHA models data transfer time using a two-component formulation: a fixed initialization cost and a variable, data-dependent cost. Empirical analysis (Fig. 4.2)

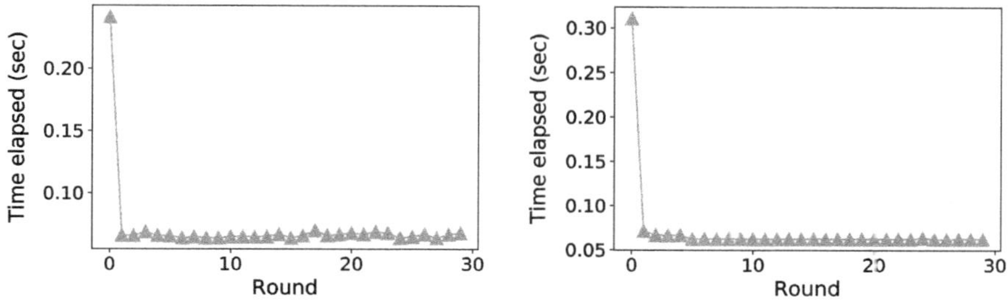

(a) M1. (b) M2.

Fig. 4.2 Communication overhead is significantly larger in the first round

reveals that the first CPU-GPU transfer incurs a substantially larger latency than subsequent transfers, indicating a significant fixed overhead inherent to the communication setup.

We therefore decompose the communication cost into (1) a fixed cost α, determined by the hardware environment and measured via controlled transfer operations (this cost is amortized across all N training batches) and (2) a variable cost, modeled similarly to the computation cost. It accounts for the transfer of both dense feature matrices (using variables n_j, m_j, f_n, and f_m) and sparse adjacency matrices (using n_j and m_j). Here, n_j and m_j denote the node and edge counts of batch $\mathcal{B}_j$, while f_n and f_m represent the dimensionalities of its node and edge features. The total communication cost for a batch $\mathcal{B}_j$ is expressed as:

$$\begin{aligned}\text{commCost}(\mathcal{B}_j) &= \frac{\alpha}{N} + Poly_d(n_j, m_j, f_n, f_m) + Poly_s(n_j, m_j) \\ &= \frac{\alpha}{N} + Poly(n_j, m_j, f_n, f_m),\end{aligned} \tag{4.3}$$

where Poly$(\cdot)$ is a polynomial function capturing the variable transfer time.

Together, the computation and communication models allow DAHA to estimate the cost of any sequence of fine-grained operations on arbitrary input data across heterogeneous devices. By learning operation- and device-specific regression parameters, the framework provides the predictive foundation necessary for optimizing pipeline parallelism during the training stage based on batch characteristics.

4.2.3 *Hybrid CPU-GPU Optimizations*

A core scheduling criterion is that the GPU should be engaged if its expected performance gain compensates for the necessary data movement. To exploit such opportunities, DAHA introduces three major hybrid CPU-GPU optimizations.

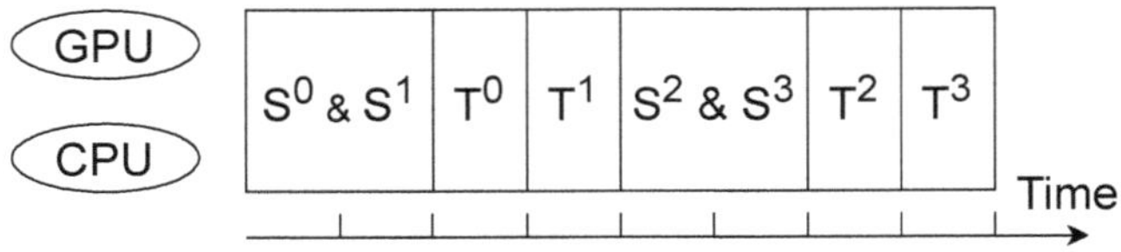

Fig. 4.3 Group-based in-turn pipeline. S^i and T^i mean the sampling and training stages of ith epoch, respectively

Group-Based Pipelining of Batch Preparation and Model Training

The conventional GNN training pipeline interleaves batch preparation (CPU-based sampling) and neural computation (GPU-based training) within a single batch stream. In contrast, DAHA introduces a group-based in-turn pipelining strategy. This approach decouples the two stages, first dedicating all available CPU and GPU resources to prepare multiple consecutive batches which may span epochs before switching to a consolidated training phase on the resulting batch groups (Fig. 4.3). During batch preparation, only subgraph structures are stored to conserve memory, omitting feature duplication. This grouping of homogeneous tasks enhances data and device locality, amortizing memory access and kernel launch overheads across batches, thereby reducing the average per-batch cost as a principle analogous to economies of scale in production. However, excessive grouping introduces scheduling complexity and strains device memory and compute capacity, eventually increasing per-unit cost and mirroring diseconomies of scale. Thus, DAHA dynamically balances these opposing effects to optimize throughput.

Effectively utilizing heterogeneous hardware for both batch preparation and neural training requires accurate predictions of sampling time and MFG statistics. However, accurate and efficient estimation is nontrivial, complicated by the stochastic nature of sampling algorithms, the complex topological structure of overlapping neighborhoods, and shuffling-induced uncertainty in batch composition. To overcome these difficulties, DAHA introduces **adaptive shuffling** and **progressive scheduling**.

Adaptive Shuffling. While the impact of shuffling is rarely examined in prior work, our empirical analysis reveals its substantial dual effect on model efficacy and training throughput. Specifically, enabling shuffling typically improves test accuracy by over 1%, enhancing generalization, but concurrently reduces end-to-end training speed by up to 10%. To navigate this trade-off and establish a reliable basis for estimating batch preparation time and MFG statistics, DAHA introduces an adaptive shuffling strategy. Inspired by transfer learning, this approach initially disables shuffling during early epochs, forming an efficient "pre-training" phase. With target nodes now deterministic, DAHA can employ predictive forecasting (via progressive scheduling) for stable estimation. Once accuracy gains diminish, indicating a suitably "pre-trained" model has been obtained, DAHA enables shuffling as a "fine-tuning" step to retain generalization benefits. To counteract the pipeline underutilization caused by shuffling uncertainty, DAHA dynamically fills idle device gaps with neural operations that exhibit favorable data locality. To determine the optimal transition point between shuffling modes, DAHA monitors the convergence trajectory and automatically identifies its elbow point where marginal accuracy improvements become negligible. This balances final model quality against runtime efficiency. After switching, only a few additional epochs with shuffling enabled are needed for convergence.

Progressive Scheduling. Under the shuffle-off mode, batch sampling time and MFG statistics become predictable, allowing DAHA to frame their estimation as a forecasting task. This formulation underpins the progressive scheduling framework of DAHA. Initially, sampling requests are distributed evenly across devices to collect baseline observations. Given the observed stability of these variables which fluctuate around a median without seasonal trends, DAHA employs a moving median method for efficient and robust forecasting. During the initial profiling phase, GPUs typically complete sampling tasks earlier than CPUs. To maintain high hardware utilization, DAHA dynamically fills GPU idle time with suitable neural operations (as in the shuffle-on mode), ensuring devices remain saturated. In subsequent rounds, with refined estimates from forecasting, DAHA schedules sampling workloads across CPU and GPU workers. We formalize this as a machine scheduling problem: given n batches with predicted processing times, where GPU sampling is uniformly faster by a factor s, the goal is to minimize makespan. This problem admits a Fully Polynomial-Time Approximation Scheme (FPTAS) with $O(n/\epsilon^2)$ complexity, guaranteeing a solution within $(1+\epsilon)$ of optimal [13]. Minimizing makespan inherently promotes balanced workloads and high utilization, as any idle device could otherwise shorten the makespan. Since sampling estimates remain static under shuffle-off, the scheduling problem is solved once and reused across all subsequent rounds.

Rewriting Rules for Intra-Batch Execution Plan Graphs

In addition to group-based in-turn pipelining, DAHA incorporates intra-batch and inter-batch neural training optimizations to better match computation to hardware and data. The decoupled design of the pipeline makes full batch statistics available beforehand, allowing DAHA to use its cost models (Eq. 4.1–4.3) for precise execution planning at both granularities each epoch. We begin with the intra-batch approach.

Abstraction as Execution Plan Graph (EPG). As noted in Sect. 4.2.1, existing approaches often impose a rigid hardware dichotomy, restricting CPUs to graph operations and GPUs to neural computations. To unlock a richer optimization space, DAHA treats both CPU and GPU as general-purpose processors capable of executing any operation. This perspective reframes GNN training around its core linear algebra components. A typical forward propagation is represented as:

$$Z^{l+1} = AH^l W^l, \ H^{l+1} = \sigma(Z^{l+1}), \tag{4.4}$$

where A is a sparse adjacency matrix, H^l is a dense node embeddings, and W^l is a dense weight matrix at layer l.

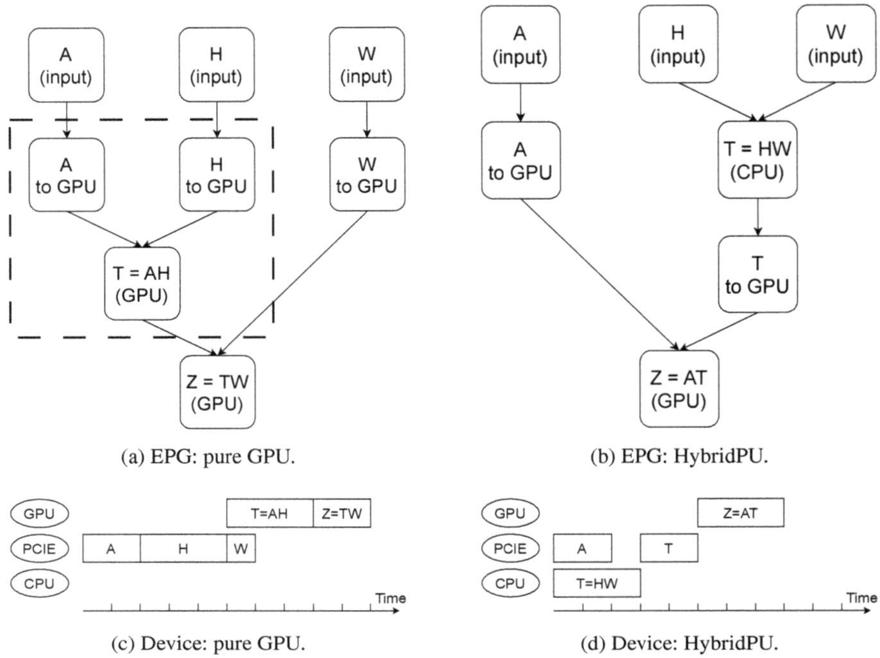

(a) EPG: pure GPU. (b) EPG: HybridPU.

(c) Device: pure GPU. (d) Device: HybridPU.

Fig. 4.4 Pure GPU versus HybridPU

To reason about scheduling, DAHA abstracts this computation as an Execution Plan Graph (EPG), a directed acyclic graph where nodes represent fine-grained operations (e.g., data transfer, spmm, mm) and edges denote data dependencies. Figure 4.4a illustrates the EPG for a classic pure-GPU execution of $Z^{l+1} = AH^l W^l$, highlighting dependencies such as the need to transfer matrices A and H to the GPU before their multiplication can begin. The corresponding per-device execution timeline, shown in Fig. 4.4c, reveals the inefficiency of this plan. Due to PCIE bandwidth contention, transfers of A and H are prioritized. Only the transfer of W can be overlapped with the subsequent sparse-dense multiplication ($T = AH$), leaving significant pipeline gaps and resulting in poor intra-batch device utilization.

Model Training with Hybrid Processing Units (HybridPU). The EPG rewriting module of DAHA, HybridPU, generates hybrid CPU-GPU execution plans to enhance parallelism. A key pre-transformation rule shifts the dense multiplication HW to the CPU, overlapping it with data transfer and reducing communication volume by sending only the result. This is highly effective for inference and forward passes. Figure 4.4b and d illustrate this pre-transformation rewriting rule. An alternative pre-aggregation rule performs sparse aggregation AH on the CPU. While usually slower for single batches, it can better utilize idle CPU resources in a multi-batch pipeline. DAHA uses its cost model and batch statistics to automatically select the most beneficial rule for a given context, an optimization applicable even to full-graph training.

Inter-Batch Scheduling Optimizations

DAHA integrates fine-grained intra-batch optimization with high-level inter-batch scheduling, improving both device utilization and overall training speed. The inter-batch scheduling problem is to schedule the operations of multiple batches, where each batch is a directed acyclic graph of dependent tasks distributed across CPU and GPU workers sharing PCIe bandwidth, using per-operation cost estimates, to minimize the overall makespan. The formulation subsumes intra-batch EPG rewriting, as operation order can be altered without violating dependencies. This allows the scheduler to inherit intra-batch optimizations directly. GPU idle time is further reduced by dynamically dispatching ready operations to available GPUs, and bounded staleness is used to increase pipeline opportunities.

This scheduling problem is a variant of the Flexible Job-Shop Problem (FJSP), which is NP-Hard [2]. The FJSP is notoriously intractable, considered among the most difficult combinatorial problems, and existing approximations often lack strong performance bounds [3, 11]. The challenge is exacerbated by the vast space of possible intra-batch execution plans that must be considered.

Algorithm Overview. To enable efficient online planning of intra- and inter-batch execution, DAHA employs a heuristic bi-level optimization strategy inspired by expectation-maximization (EM) and grouping principles (Algorithm 4) [9]. The core insight is that, due to finite hardware resources and task-switching overhead, only a limited number of batches can be executed in a tightly interleaved manner. Consequently, batches are partitioned into groups, with the group size determined by the available workers. Optimizing within these smaller groups significantly reduces the search space.

Formally, let $\mathcal{G} = \{\mathcal{G}_i\}$ be a partition of the batch set $\mathcal{B}$. The strategy proceeds in two levels. At the **lower level**, given a grouping $\mathcal{G} = \{\mathcal{G}_i\}$, DAHA optimizes the execution plan for each group $\mathcal{G}_i$, leveraging the valid EPG rewriting rules from its intra-batch strategies. At the **upper level**, DAHA evaluates the pipeline efficiency of the optimal plans for each group. It then refines the grouping itself, particularly for inefficient groups, by regrouping their batches. These new groups are fed back to the lower level for re-optimization.

This iterative, feedback-driven process between grouping refinement and intra-group optimization resembles an EM algorithm. Full details are provided in Algorithm 4. For the algorithm complexity, the per-iteration complexity in the upper level is dominated by the grouping update which takes $O(n \log n)$ time, involving sorting and efficient neighbor searches. For the lower level, with the group size β bounded by the number of available workers, the lower-level complexity is capped at $O(\beta! \cdot n)$. Given typical values of $\beta < 10$ and $n < 100$, the total per-iteration complexity of $O(n \log n + \beta! \cdot n)$ is computationally manageable.

Algorithm 4 Bi-level EM-based Scheduling Algorithm.

Require: $\mathcal{M} = \{\mathcal{M}_1, ..., \mathcal{M}_m\}$: available processing units; $\mathcal{B} = \{\mathcal{B}_1, ..., \mathcal{B}_n\}$: batches to schedule; $\{\mathcal{R}_1, ..., \mathcal{R}_r\}$: EPG rewriting rules; $compCost_{M_i}(\cdot)$ and $commCost(\cdot)$: computation and communication cost models; μ^*: a parameter in (0, 1) controlling the grouping quality; α: a rate controlling the increase of grouping quality; β: maximum size of a batch group; τ: maximum iterations.
Ensure: An execution plan of the batches on the processing units.

Initiate a grouping $\mathcal{G} = \{\mathcal{G}_i\}$ randomly that partitions $\mathcal{B}$
$f \leftarrow$ a list of zeros with length $|\mathcal{G}|$
$\mu, iterCount \leftarrow 0$
while $iterCount < \tau$ and $\mu < \mu^*$ **do**
    $\mu \leftarrow \mu + \alpha; iterCount \leftarrow iterCount + 1$
    for all $\mathcal{G}_i \in \mathcal{G}$ **do**
        $f[i] \leftarrow 0$
        Apply the candidate EPG rewriting rules $\{\mathcal{R}_1, ..., \mathcal{R}_r\}$ to $\mathcal{G}_i$ to list the feasible execution plans of the group
        Apply $compCost_{M_i}(\cdot)$ and $commCost(\cdot)$ to estimate the makespan of each execution plan
        Determine the execution plan $\mathcal{P}_i$ that minimizes the makespan and the minimum makespan C_i
        if $C_i > (1-\mu) \cdot \sum_{\mathcal{B}_j \in \mathcal{G}_i} cost_{\mathcal{P}_i}(\mathcal{B}_j)$ **then**
            $f[i] \leftarrow 1$
        end if
    end for
    $\Delta\mathcal{B} \leftarrow \{\mathcal{B}_j : \mathcal{B}_j \in \mathcal{G}_i \ and \ f[i] = 1\}$
    for all $\mathcal{B}_j \in \Delta\mathcal{B}$ **do**
        $k \leftarrow \arg\min_i cost_{R_i}(\mathcal{B}_j)$
        $y_j \leftarrow \frac{compCost_{R_k}(\mathcal{B}_j)}{commCost_{R_k}(\mathcal{B}_j)}$
    end for
    $\Delta\mathcal{G}^* \leftarrow \arg\min_{\Delta\mathcal{G}} \max_{\Delta\mathcal{G}_i} |\log(\prod_{\mathcal{B}_j \in \Delta\mathcal{G}_i} y_j)|$ under the constraint that $|\Delta\mathcal{G}_i| \leq \beta$ for $\forall \Delta\mathcal{G}_i \in \Delta\mathcal{G}$, where $\Delta\mathcal{G} = \{\Delta\mathcal{G}_i\}$ is a grouping that partitions $\Delta\mathcal{B}$ and $|\Delta\mathcal{G}_i|$ is the number of batches in the group $\Delta\mathcal{G}_i$
    $\mathcal{G} \leftarrow \{\mathcal{G}_i : f[i] = 0\} \cup \Delta\mathcal{G}^*$
end while
Determine the execution plan $\mathcal{P}_i$ that minimizes the makespan for each $\mathcal{G}_i$
return $\mathcal{G} = \{\mathcal{G}_i\}$ and $\{\mathcal{P}_i\}$

4.2.4 Experiments

This section covers the evaluation of the effectiveness and efficiency of DAHA. We first elaborate on the experimental setups.

Setup. Experiments are conducted on two hardware configurations. The default machine (M1) features an Intel i9-12900KF CPU, 32 GB DDR5 RAM, and an NVIDIA RTX 3080 Ti GPU (12 GB), running PyTorch v1.12, DGL v0.9, and CUDA v11.6. To assess hardware adaptability, we also use a machine (M2) with an Intel Xeon Gold 6240 CPU, an NVIDIA RTX 2080 Ti GPU, and CUDA v10.1.

Table 4.1 Dataset statistics

Dataset	#Nodes	#Edges	#Feat.	#Class	Vol_G	Vol_F
Cora	2708	10556	1433	7	0.403	14.803
Citeseer	3327	9104	3703	6	0.350	46.997
Pubmed	19717	88648	500	3	3.382	38.000
CS	18333	163788	6805	15	3.125	476.000
Physics	34493	495924	8415	5	9.459	1108.000
Arxiv	169343	1166243	128	40	44.898	82.687
Products	2449029	247436304	100	47	4720.000	934.230
Reddit	232965	229231784	602	41	4372.450	534.992

"#" means the number of. Vol_G and Vol_F are the size of the sparse adjacency matrix and dense feature matrix in MB, respectively

Datasets. The evaluation employs standard benchmark datasets (see Table 4.1) covering citation networks (Cora, Citeseer, Pubmed, ogbn-arxiv), co-author networks (CS, Physics), a co-purchase graph (ogbn-products), and a social graph (Reddit). All are publicly available through DGL [26], OGB [14], and PyG [8]. This diverse selection, varying in size, density, and feature dimensions, to test the capacity of DAHA to optimize training across different data types.

GNN Models. The evaluation employs three widely used message-passing GNN architectures: the Graph Convolutional Network (GCN) [19] trained on the full graph; GraphSAGE [12] trained with node-wise mini-batch sampling; and GraphSAINT [33] trained with subgraph-level mini-batch sampling. All models are trained for 100 epochs. For the mini-batch models, we set the batch size to 8192; for the node sampler of GraphSAINT, a budget of 16384 nodes is used.

Baselines. The performance of DAHA is evaluated against several prominent GNN training frameworks and systems: PyTorch [22], DGL [26], DGL+ (a pure-GPU variant of DGL), ByteGNN [35], and SALIENT [16]. As DAHA is implemented on top of PyTorch, its performance gain is benchmarked relative to the base PyTorch execution. DGL serves as the widely-adopted baseline for mini-batch GNN training, with DGL representing its classic CPU-sampling and GPU-training mode and DGL+ representing its end-to-end GPU mode. The original ByteGNN system is designed for CPU-only execution. For a fair hardware comparison, it is paired with DGL+ on the GPU, effectively utilizing all available resources. SALIENT is a recent mini-batch training system notable for its pipeline parallelism and optimized sampling. For consistency, the DGL sampler is used across all baselines. GCN models are tested on smaller datasets (Cora, Citeseer, Pubmed, CS, Physics) to evaluate full-graph training, while GraphSAGE and GraphSAINT models are tested on larger datasets (Arxiv, Products, Reddit) to evaluate mini-batch training.

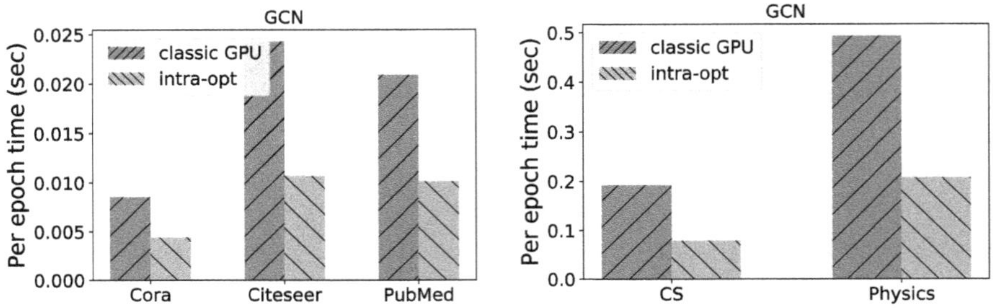

(a) DGL datasets. (b) PyG datasets.

Fig. 4.5 Overall performance for GCN

Overall Performance

The overall system performance is evaluated by measuring the average training time per epoch over 100 epochs (evaluation time excluded). On full-graph GCN tasks (Cora, Citeseer, Pubmed, CS), the computational speedup from the EPG rewriting module of DAHA is verified. For mini-batch tasks with GraphSAGE and GraphSAINT (Physics, Arxiv, Products, Reddit), DAHA is compared against all frameworks to benchmark its end-to-end pipeline efficiency. The corresponding model accuracy is also reported to confirm that speedups are not achieved at the expense of predictive performance.

GCN Speedup. Figure 4.5 compares per-epoch training times for GCN, showing DAHA consistently outperforming PyTorch on five diverse datasets, with end-to-end speedups ranging from 48.4 to 59.1%. As full-graph GCN training excludes sampling overhead, these gains are attributed purely to the execution plan optimizations of DAHA (e.g., pre-transformation via EPG rewriting). This shows DAHA can effectively leverage hybrid CPU-GPU computation by utilizing otherwise idle hardware. The consistent high speedup across datasets of varying size, density, and feature dimension confirms the adaptability of DAHA in generating efficient, data-aware execution plans.

Mini-batch Training Speedup. DAHA demonstrates significant and consistent acceleration for GraphSAGE and GraphSAINT across the Arxiv, Products, and Reddit datasets (Fig. 4.6), achieving up to 81% faster training than DGL. It also outperforms the pipelined system SALIENT by up to 73%. This improvement stems from the dual-advantage design of DAHA. The first is its group-based pipeline reduces effective per-epoch sampling time, and the second is its inter-batch planning exploits deeper optimization opportunities, yielding faster computation. While ByteGNN is competitive in some scenarios, it lacks the coordinated, inter-batch optimization that defines the advantage of DAHA. Across all workloads, DAHA consistently achieves the fastest end-to-end time, validating its adaptability to diverse data and model specifications.

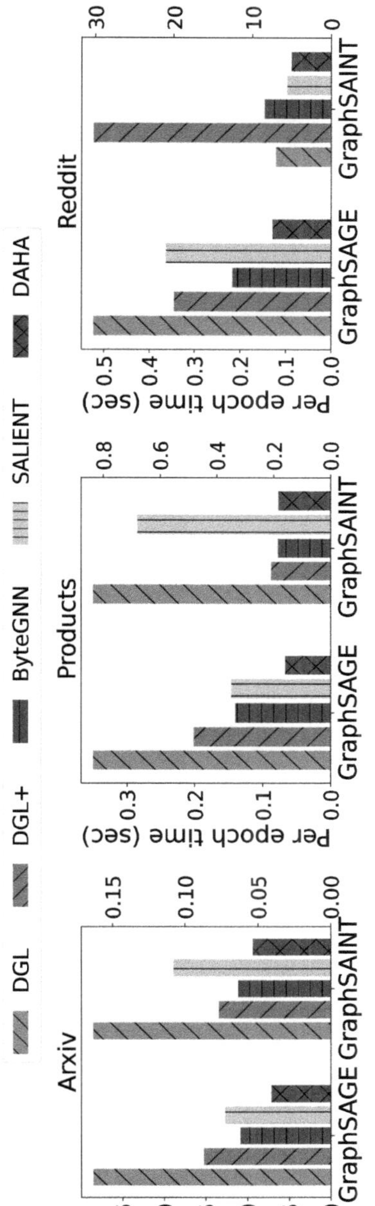

(a) Arxiv. (b) Products. (c) Reddit.

Fig. 4.6 Overall performance for mini-batch. The left y-axis is for GraphSAGE and the right y-axis is for GraphSAINT

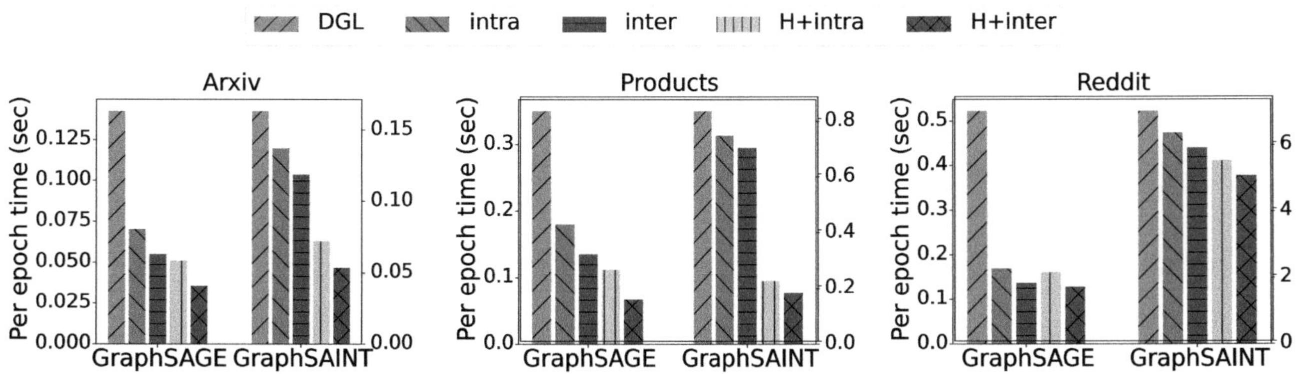

(a) Arxiv. (b) Products. (c) Reddit.

Fig. 4.7 Ablation study for mini-batch. Left y-axis is for GraphSAGE and right y-axis is for GraphSAINT

Table 4.2 GraphSAGE accuracy with various shuffling mode

Dataset	Shuffle-on	Shuffle-off	Adaptive shuffle
Arxiv	69.58	68.11	69.21
Products	76.76	76.40	76.50
Reddit	96.29	96.22	96.27

Mini-batch Accuracy. Table 4.2 compares test accuracy under three shuffling modes: shuffle-on (standard), shuffle-off, and adaptive shuffling. Shuffle-on yields the best accuracy, while shuffle-off degrades generalization, incurring up to a 1.5% drop on Arxiv. The adaptive strategy of DAHA mitigates this by pre-training with shuffle-off before enabling shuffling for fine-tuning, achieving accuracy on par with shuffle-on and substantially outperforming shuffle-off.

Breakdown Analysis

DAHA Variants. To understand the sources of the performance gains of DAHA, its three key components, (1) group-based in-turn pipelining, (2) intra-batch EPG rewriting, and (3) inter-batch scheduling, are isolated and evaluated. The predictive accuracy of its data-aware and hardware-aware cost model is also examined. Four variants of DAHA are introduced below.

- "intra": DGL CPU sampler with intra-batch optimizations;
- "inter": DGL CPU sampler with both intra- and inter-batch optimizations;
- "H+intra": group-based in-turn pipelining with intra-batch optimizations;
- "H+inter": group-based in-turn pipelining with both intra- and inter-batch optimizations (the default variant of DAHA).

The training time statistics for GraphSAGE and GraphSAINT (Arxiv, Products, Reddit) are detailed in Table 4.3. Combined with Fig. 4.7, these results provide an ablation analysis, showing that (1) intra-batch EPG rewriting speeds up training compared to DGL (intra is better than DGL); (2) inter-batch scheduling improves upon intra-batch optimization (inter is better than intra); and (3) group-based in-turn pipelining further enhances efficiency (H+intra is better than intra, and H+inter is better than inter).

The Effects of the Cost Model. The efficiency and accuracy of the cost model are validated in Tables 4.4 and 4.5. The model, fitted on 10 Pubmed subgraphs, accurately predicts operation costs across datasets and devices, including data transfers. It correctly captures the greater GPU speedup for spmm versus mm, validating the pre-transformation strategy. This one-time per-machine calibration is fast, adds negligible runtime overhead (0.005 s per-100 predictions), and generalizes well, demonstrating the data and hardware awareness of DAHA.

Table 4.3 Breakdown of DAHA modules and comparison with others on mini-batch training

Dataset	GNN	sample	intraT	interT	DGL	DGL+	ByteGNN	SALIENT	intra	inter	H+intra	H+inter
Arxiv	GraphSAGE	4.40	2.63	1.10	14.26	7.60	5.43	6.34	7.03	5.50	5.10	3.57
Arxiv	GraphSAINT	10.24	3.47	1.61	16.32	7.68	6.36	10.79	13.71	11.85	7.21	5.35
Products	GraphSAGE	10.38	7.63	3.15	35.00	20.11	14.03	14.63	18.00	13.52	11.19	6.71
Products	GraphSAINT	67.38	7.45	3.15	83.60	20.95	18.59	68.06	74.83	70.52	22.70	18.39
Reddit	GraphSAGE	8.14	8.82	5.57	52.26	34.41	21.66	36.37	16.96	13.71	16.13	12.87
Reddit	GraphSAINT	551.59	85.81	40.81	702.74	3024.34	846.21	561.64	637.40	592.40	552.36	507.36

The records here are per-100 epoch. "sample" means the sampling time given the default DGL sampler. "intraT" and "interT" mean the computation time (training stage only, excluding sampling stage) using the intra-batch and inter-batch optimizations of DAHA, respectively. "DGL" reports CPU-sample-GPU-train time and "DGL+" reports pure-GPU time of DGL. The last four columns are variants of DAHA

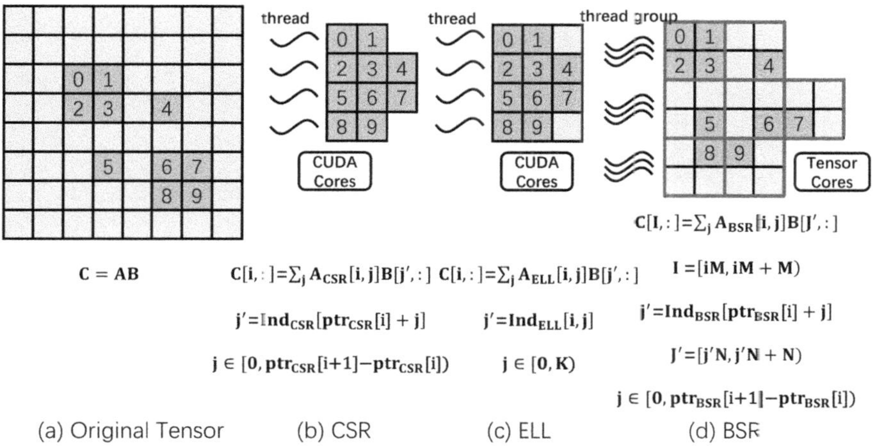

Fig. 4.8 Common sparse formats and corresponding SpMM implementations. The non-zeros are colored. For CSR, $\mathrm{Ind}_{\mathrm{CSR}}[\mathrm{ptr}_{\mathrm{CSR}}[i] : \mathrm{ptr}_{\mathrm{CSR}}[i+1]]$ stores the column indices of the non-zeros in row i. For ELL, $\mathrm{Ind}_{\mathrm{ELL}}[iK : iK + K]$ stores the column indices of the non-zeros and the padded zeros in row i, where K is the maximum number of non-zeros per row. For BSR, $\mathrm{Ind}_{\mathrm{BSR}}[\mathrm{ptr}_{\mathrm{BSR}}[i] : \mathrm{ptr}_{\mathrm{BSR}}[i+1]]$ stores the column indices of the non-zero blocks in block row i; (M, N) is the block size, e.g., (M, N) is (2, 2) in (d)

Table 4.4 Preprocessing time in seconds for cost model

Machine	Data collection	Regression
M1	37.8564	0.0205
M2	64.2065	0.2793

4.3 Hybrid Sparse Formats for Sparse Operators

4.3.1 Motivation

As mentioned in Sect. 4.1, existing sparse formats have advantages and disadvantages when dealing with different sparsity patterns. For example, the CSR format compactly stores the non-zeros of a sparse matrix by rows, as Fig. 4.8b shows. In SpMM, it is common to make each GPU thread compute one row of the output ($C[i, :]$ in Fig. 4.8b), but the irregular row lengths after compression will cause workload imbalance among threads which damages the operator efficiency. The ELL format is different from CSR in that it pads zeros to each row of the ragged compressed matrix and therefore returns a regular matrix (row length is K in Fig. 4.8c). It removes the ptr array in CSR to simplify the process of getting valid column indices, eliminates control flow divergence in threads, and enables memory coalescing if with column-major storage, allowing better operator efficiency. The BSR format stores non-zero blocks (i.e., the blocks with at least one non-zero), e.g., Fig. 4.8d stores

Table 4.5 Comparison of true and predicted time cost (in seconds)

Machine	Operation	spmm						mm						Data transfer		
	Dataset	Pubmed		CS		Physics		Pubmed		CS		Physics		Pubmed	CS	Pyhsics
	Device	CPU	GPU	CPU	GPU	CPU	GPU	CPU	GPU	CPU	GPU	CPU	GPU	PCIE	PCIE	PCIE
M1	Truth	5.8000	0.5000	68.8993	6.2001	243.9285	17.3027	0.3001	0.1000	1.0000	0.1001	2.4999	0.2000	0.6000	6.0110	13.8314
	Prediction	5.7397	0.4994	69.2511	5.5836	235.2394	17.6340	0.3061	0.0945	0.9029	0.0988	2.4201	0.2044	0.5434	5.9200	13.6227
M2	Truth	4.8111	0.2213	57.3462	1.8583	231.3737	6.1514	0.0209	0.0147	0.8864	0.2030	2.2079	0.3689	0.9971	10.9417	25.2243
	Prediction	4.8857	0.2224	58.1973	1.7921	276.5741	6.3681	0.0216	0.0142	0.9073	0.1890	2.2909	0.3719	0.9793	11.0075	22.1922

The statistics are per-100 epoch. Data transfer involves both the sparse adjacency matrix and the dense feature matrix

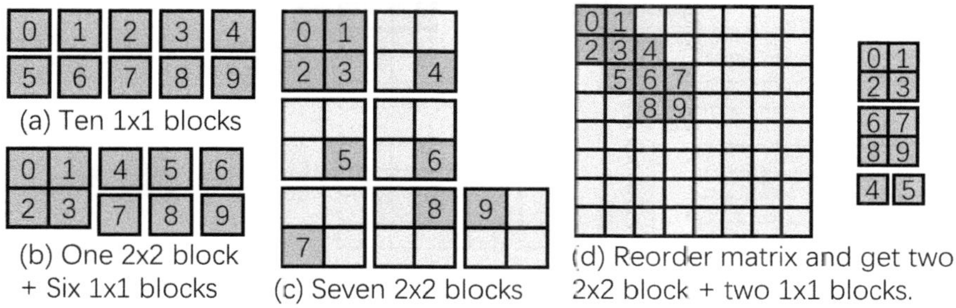

Fig. 4.9 Examples of block collections to cover non-zeros

non-zero 2×2 blocks. Compared with CSR, BSR stores the column indices for blocks instead of those for non-zeros (Ind_{BSR}, ptr_{BSR} in Fig. 4.8d) and hence reduces the memory usage for sparsity structure information. It also provides opportunities for vectorization or hardware intrinsics [46]. For instance, in SpMM, it is common to use a group of threads (a GPU thread block) to compute a set of output rows like $C[I, :]$ in Fig. 4.8d, which can run on the powerful GPU Tensor Cores for higher efficiency. However, the extra computation workload brought by padding may be a severe problem for both ELL and BSR. Therefore, ELL is more suitable for matrices where the number of non-zeros per row does not vary a lot, and BSR is more suitable for matrices with dense sub-matrices. It is well-recognized that no single format can suit all sparse tensors. Moreover, if a sparse tensor has different sparsity patterns in different regions, we may need to combine different formats together to achieve optimal performance.

To deal with the diverse sparsity patterns of different sparse tensors and of different parts of a sparse tensor, people try to use hybrid formats, i.e., taking the sparse operator into consideration, they decompose a sparse tensor into several sparse sub-tensors and select the best sparse format for each of them [48, 55, 65, 66, 76, 78], then they schedule the computation based on the hybrid format.

There are two major types of decomposition methods: the set-cover-based decomposition [78], where the decomposed sparse sub-tensors may have intersections, and the partition-based decomposition [48, 55, 65, 66, 76], where there are no intersections between the decomposed sparse sub-tensors. However, the existing works often make a trade-off between search space and search time. As a result, their search spaces are limited in some cases, limiting the optimized operator efficiency they can achieve.

Set-Cover-Based Decomposition. Given a sparse tensor, a sub-tensor, also called a block, covers a set of non-zeros of the original tensor. The set-cover-based decomposition is to find the best collection of blocks such that all the non-zeros of the original sparse tensor are covered to ensure the accuracy of the computation result of the sparse operator. The existing work in this category is [78]. It only considers using the block format to store the decomposed blocks, i.e., if not full of zeros, each block will be stored entirely in the memory, including both the zeros and the non-zeros in it.

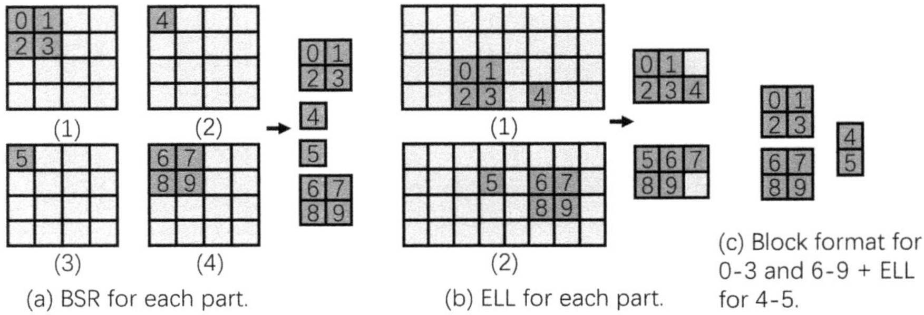

Fig. 4.10 **a, b** Two examples of partition-based decomposition; **c** an example of multi-level decomposition

We use the sparse tensor in Fig. 4.8a to explain this kind of decomposition below. In practice, the complete computation of a sparse operator will be divided into several sub-tasks and assigned to multiple parallel computation units. For simplicity, in this example and other examples in Sect. 4.3.1, we assume (1) each sub-task runs the computation related to one block, and hence we directly refer to the sub-task cost (i.e., the running time) as the corresponding block cost; (2) sub-tasks run on one computation unit so the total cost of the selected blocks is the summation of their costs. As the candidate blocks are generated by partitioning the sparse tensor evenly in [78], we assume the block shapes we consider are 1×1, 2×2, and 4×4 and the associated costs are 2, 4, and 15, respectively. Figure 4.9 shows some possible block collections to cover the non-zeros. If we only select 1×1 blocks to cover the non-zeros (Fig. 4.9a), the total cost is 20 (10×2). If we only select 2×2 blocks, 7 blocks will be selected in the end (Fig. 4.9c), the total cost is 28 (7×4). The example of Fig. 4.9b considers hybrid block shapes: it selects one 2×2 block for the non-zeros 0–3 and six 1×1 blocks for the remaining non-zeros, and the total cost is 16 ($4 + 6 \times 2$). However, as [78] does not consider other possible sparse formats for the local sparse patterns other than the block format and does not perform the possible transformations of the original sparse tensor, like matrix row/column reordering, its search space of the decomposition solutions is limited. For example, as shown by Fig. 4.9d, we can first transform the sparse matrix by reordering its rows from rows 0, 1, 2, 3, 4, 5, 6, 7 to rows 2, 3, 5, 6, 0, 1, 4, 7, and reordering its columns from columns 0, 1, 2, 3, 4, 5, 6, 7 to columns 2, 3, 5, 6, 0, 1, 4, 7. Then we use two 2×2 blocks to cover non-zeros 0–3 and 6–7, respectively, and another two 1×1 blocks for non-zeros 4, 5. The cost is 12 ($2 \times 4 + 2 \times 2$).

Partition-Based Decomposition. Given a sparse matrix, the partition-based decomposition method will partition the matrix into several parts such that we can find a good sparse format for each part that is suitable for the corresponding local sparse pattern. To make the local sparse patterns more obvious, the existing works [48, 55, 76] will transform the matrix, like reordering the rows by the number of non-zeros. Taking the sparse matrix of Fig. 4.8a as an example, Fig. 4.10a, b show some partition-based decomposition results. Figure 4.10a divides the original tensor into four parts and reorders the rows and columns of each part. Each part is then stored in the BSR format. The total cost is 12 ($2 \times 4 + 2 \times 2$). Figure 4.10b partitions the original tensor into two parts and stores each part in the ELL format. Similar to the block cost, we assume each of the ELL-format sub-matrices is of cost 6, then the total cost is 12 (6×2). However, the search space of the partition-based decomposition method is also limited because there may be some non-zeros in a submatrix that do not fit the selected format, and it will be better for us to use another format to store such non-zeros of each submatrix. For example, if we select two ELL formats as in Fig. 4.10b, we need to pad two zeros, which wastes the computation. By contrast, as shown by Fig. 4.10c, we can select the block format for "0–3" in the first matrix part ("4" is left) and the block format for "6–9" in the second part ("5" is left). Then we use ELL for the remaining "4" and "5". Assume the cost of the ELL format for 4 and 5 is 3. The total cost is 11 ($2 \times 4 + 3$). Our experiments show that considering such multiple levels (two levels in this example) of format selection can lead to 3.6$\times$ operator efficiency improvement in some cases.

From the above examples, we observe that to find a good decomposition of a given sparse tensor, it is necessary to make the search space large enough. Specifically, we should not only extend the breadth of the search space by considering the matrix transformations and various sparse formats (as shown by Figs. 4.9d and 4.10c) but also extend the depth of the search space by considering the multi-level decomposition (i.e., select formats for a part of the non-zeros first, then reorganize the remaining non-zeros and repeat this process, as shown by Fig. 4.10c, where the two block formats are selected for the first level, and the ELL format is selected for the second level). Therefore, we face two major challenges in finding a good sparse format decomposition: (1) how to construct a large search space with its breadth and depth extended; (2) how to search efficiently in the large search space to find a good hybrid sparse format.

Before delving into the proposed solution, we first provide several preliminary definitions. A sparse tensor is a tensor where only a part of the elements are non-zeros, and a sparse format is how we store the non-zeros.

Definition 4.1 (Sparse Tile) A sparse tile t is a sub-tensor of a sparse tensor with a specified sparse format, denoted by (E, α), where E is the set of non-zeros covered by t and α is the sparse format to store E.

For example, the first sparse tile in Fig. 4.10a can be denoted by ($E = \{(2, 2), (2, 3), (3, 2), (3, 3)\}$, $\alpha =$ block format), where E stores the coordinates of the four non-zeros in the original sparse matrix. Given a sparse tensor, denoted by T, we can generate multiple candidate sparse tiles by transforming T (the transformation details will be introduced later), which are first-level sparse tiles. After we select one or more sparse tiles, we can get a new sparse tensor containing the remaining non-zeros of T, and we can transform the new tensor to get new sparse tiles, i.e., the second-level sparse tiles. By repeating this process, our search space includes multi-level sparse tiles. Taking the sparse matrix in Fig. 4.8a and the sparse tiles

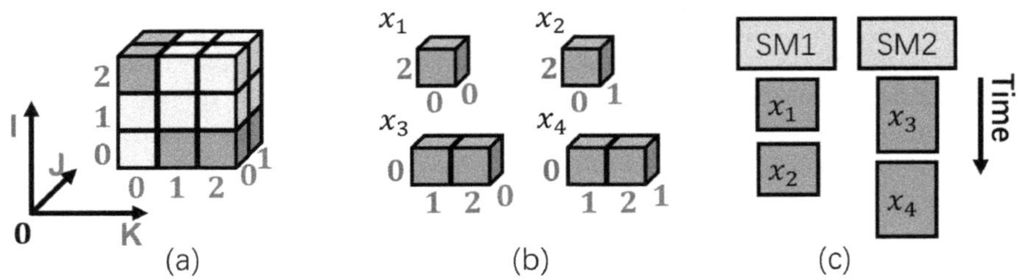

Fig. 4.11 **a** The iteration space of an SPMM operator O is three-dimensional (I, J, K), where the gray points can be skipped while the yellow points cannot; U includes the yellow points. **b** By selecting two sparse tiles $t_1 = (\{(2, 0)\}, \text{ELL})$, $t_2 = (\{(0, 1), (0, 2)\}, \text{ELL})$ for the sparse input tensor and dividing loop J into $J_1 = \{0\}$, $J_2 = \{1\}$, we get four sub-tasks of O, each covers a subset of U, e.g., x_1 has $u_1 = \{(2, 0, 0)\}$. **c** The cost of running the four sub-tasks is the maximum cost of the SMs, i.e., the cost of SM2

in Fig. 4.10c as an example. The sparse tiles storing $0 - 3$ and $6 - 9$, respectively, are first-level sparse tiles because they can be obtained by transforming the original sparse matrix in Fig. 4.8a; while the sparse tile with ELL format storing $4, 5$ is a second-level sparse tile because it can only be generated after selecting the previous two sparse tiles and reorganizing the remaining non-zeros.

We need to select a collection of sparse tiles to determine the hybrid sparse format (a sparse format decomposition) for T, which is defined below.

Definition 4.2 **(Valid Tile Selection)** A valid tile selection for a sparse tensor T, which is denoted by S, is a set of sparse tiles such that all non-zeros of T are covered by at least one tile in S.

Given an operator and the tile selection(s) for its sparse input tensor(s), we can partition the computation task accordingly into a set of sub-tasks and generate efficient code for the sub-tasks to get the final program. Specifically, the computation task of a sparse operator O can be represented by an iteration space whose dimensions correspond to the loops in the computation expression of O, and some points in the iteration space will be skipped because of the sparsity. A sub-task can be defined as follows:

Definition 4.3 **(Sub-task)** Given a sparse operator, O, and the set of points that O cannot skip in the iteration space, denoted by U, a sub-task x iterates in a sub-space of the iteration space of O and covers a subset of U, denoted by u.

Definition 4.4 **(Task Partition)** A task partition X for a sparse operator O is a set of the sub-tasks s.t. $\bigcup_{x_i \in X} u_i = U$.

Take an SpMM operator O as an example, which performs the multiplication between a sparse matrix $A_{I \times K}$ and a dense matrix $B_{K \times J}$, i.e., it computes $C[i, j] = \sum_{k \in K} A[i, k] \times B[k, j], i \in I, j \in J$. The expression includes three loops I, J, K where I, J are spatial loops and K is the reduction loop, so the iteration space of

O is 3-dimensional, but the iteration over I, K will skip some points because of the sparsity of A, while the iteration over J cannot skip any points, i.e., J is a dense loop. Figure 4.11a illustrates this iteration space. Suppose the tile selection for A is $S = \{t_1, t_2\}$, where $t_1 = (\{(2, 0)\}, \text{ELL}), t_2 = (\{(0, 1), (0, 2)\}, \text{ELL})$, the computation task may be partitioned by dividing the iteration over J into two parts $J_1 = \{0\}$, $J_2 = \{1\}$ and making each sub-task iterate over the points determined by t_i and J_j, where $t_i \in S$, $J_j \in \{J_1, J_2\}$. Therefore, we get four sub-tasks: x_1 with $u_1 = \{(2, 0, 0)\}$, x_2 with $u_2 = \{(2, 1, 0)\}$, x_3 with $u_3 = \{(0, 0, 1), (0, 0, 2)\}$, x_4 with $u_4 = \{(0, 1, 1), (0, 1, 2)\}$, which is shown in Fig. 4.11b.

Definition 4.5 (Task Partition Way) A task partition way for a sparse operator O, denoted by $p = (i_I, i_J, i_K, ...)$, is the way to partition each dense loop of O, where i_I (or $i_J, i_K, ...$) is the size of each part of the dense loop I (or J, K, ...) after partition.

In the above example, the task partition way is $p = (i_J), i_J = 1$. For a sparse operator, given a tile selection S and a task partition way p, the task partition X is determined.

The cost of a sub-task is its processing time on hardware; the cost of a task partition is the total running time to finish all the sub-tasks on hardware. Specifically, on GPU, which we target in this work, the sub-tasks will be assigned to multiple streaming multiprocessors (SMs) which work in parallel, and therefore the total running time of all the sub-tasks is the maximum running time of the SMs. Figure 4.11c shows an example of running multiple sub-tasks on SMs, in which the total cost is the cost of SM2. As we do not know how GPU will schedule the sub-tasks, we assume it will use the best schedule which minimizes the maximum running time of the SMs. The cost of a task partition is defined below.

Definition 4.6 (Cost of Task Partition) Given a task partition X with sub-tasks $x_1, ..., x_m$ and a GPU with SMs $W_1, ..., W_n$, when the assignment of the sub-tasks to the SMs is $f = \{X_1, ..., X_n\}$ (X_i is the set of the sub-tasks assigned to W_i), the cost of X under f is $\text{cost}_f(X) = \max_i \sum_{x \in X_i} \text{cost}(x)$. Therefore, the cost of X is $\text{cost}(X) = \min_f \text{cost}_f(X)$.

We now define the multi-level sparse tiles to extend the depth of the search space.

Definition 4.7 (Multi-level Sparse Tile) Given a sparse operator whose sparse input is denoted by T_0, we can get the set of possible first-level sparse tiles, denoted by ST_1, by transforming T_0; after selecting a subset S_1 of ST_1, we get a new sparse tensor T_1 with the remaining non-zeros of T_0 and a set of possible second-level sparse tiles, denoted by ST_2, by transforming T_1. We repeat this process: in the ith round, after selecting a subset S_i of ST_i, we get a sparse tensor T_i with the remaining non-zeros of T_{i-1} and a set of possible $(i + 1)$th-level sparse tiles ST_{i+1}, by transforming T_i. The process stops in round N when $S = \bigcup_1^N S_i$ is a valid tile selection for T_0.

Now, we formally define the multi-level sparse format decomposition problem (shorted for MSFD) below.

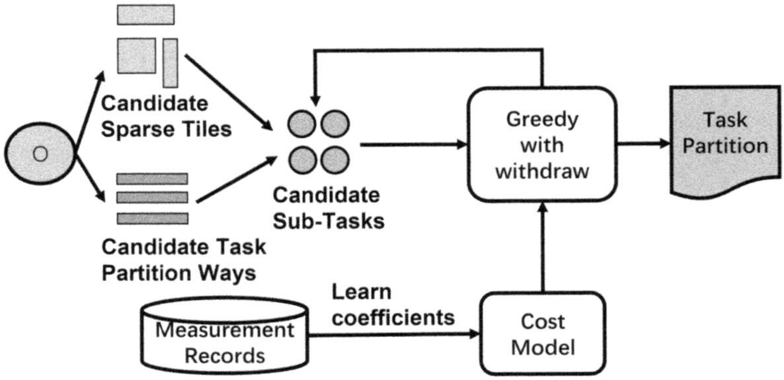

Fig. 4.12 Framework overview

Definition 4.8 (MSFD Problem) Given a sparse operator whose sparse input is denoted by T_0 and a set of task partition ways P, The MSFD problem is to get a valid tile selection S by selecting sparse tiles level by level in the way defined in Definition 4.7 and choose a task partition way $p \in P$ s.t. the cost of the corresponding task partition X, i.e., $cost(X)$, is minimized.

Theorem 4.1 *The MSFD problem is NP-hard.*

Proof sketch Consider a simplified version of the problem: we only consider the first-level sparse tiles and the GPU only has one SM. In this setting, we get a fixed set of possible sub-tasks. Finding a task partition with the minimum cost is equivalent to solving the weighted set cover problem, which is NP-hard. Therefore, the multi-level sparse format decomposition problem is NP-hard (Table 4.6). □

4.3.2 An Automatic Hybrid Sparse Format Search Framework

We propose STile to solve the MSFD problem. Figure 4.12 shows the overview of STile. Specifically, given a sparse operator O and its input tensors, we design the search spaces of possible sparse tiles and possible task partition ways (Sect. 4.3.2), which determine the search space of sub-tasks. We use a cost model to estimate the cost of a sub-task (Sect. 4.3.2), and based on it, we use a greedy-with-withdraw algorithm to get a task partition for O (Sect. 4.3.2).

Search Space

As we have mentioned, the search space of sub-tasks is determined by two factors: multi-level sparse tiles and task partition ways. Table 4.7 shows the primitives we use

Table 4.6 Notation table

Symbol	Meaning
E	A set of non-zeros
α	A sparse format
$t = (E, \alpha)$	A sparse tile
T, A, B, C	A (sparse) tensor
O	A (sparse) operator
$S = \{t_1, ..., t_m\}$	A valid tile selection
$I, J, K, ...$	The operator iteration space dimensions
U	The set of cannot-skip iteration points of O
u	A subset of U
x	A sub-task to complete a sparse operator
$X = \{x_1, ..., x_m\}$	A task partition of a sparse operator
p	A task partition way
P	A set of task partition ways
W	A compute worker (e.g., an SM for GPUs)
cost(X)	The cost of a task partition for an operator

to generate sparse tiles, which are summarized from existing works. For example, the ELL format (Fig. 4.8c) can be obtained by primitive 3 $\rightarrow$ primitive 5 (for each row). The BSR format (Fig. 4.8d) can be obtained by primitive 2 (along rows, $i = 2$) $\rightarrow$ primitive 2 (along columns, $i = 2$). The condense format in [72] can be obtained by primitive 2 (along rows, $i = 16$) $\rightarrow$ primitive 1 (along columns) $\rightarrow$ primitive 2 (along columns, $i = 16$). The hyb format in [76] can be obtained by primitive 2 (along columns, i = any value) $\rightarrow$ primitive 1 (along rows) $\rightarrow$ primitive 3 $\rightarrow$ primitive 4 ($j = 2^k$) $\rightarrow$ primitive 5 (for each row) $\rightarrow$ primitive 2 (along rows, $i = 2, 4, 8, ...$). The 1D format used for 1D tiling in [49, 77] (i.e., storing the non-zeros consecutively) can be obtained by primitive 3 $\rightarrow$ primitive 6 $\rightarrow$ primitive 4 ($j = 2^k$) $\rightarrow$ primitive 2 (along rows, $i = 1$).

In this work, we consider three basic formats for sparse tiles: the block format, the ELL format, the 1D format. If a sparse tile is obtained without applying the "Squeeze" primitive (i.e., the zeros are kept), we use the block format for it; if a sparse tile is obtained by using the "flatten" primitive, the "squeeze" primitive must also be applied to it, and this means the non-zeros are consecutively stored, so we use the 1D format for it; otherwise, we always apply the "Zero padding" primitive after "Squeeze" and use the ELL format for it. We run the computation related to block-format sparse tiles on GPU Tensor Cores, so the sub-tensor covered by such a sparse tile is required to be of certain shapes, e.g., (16, 16), (8, 16), which are consistent with the hardware intrinsics like wmma. The computation related to sparse tiles of other formats is run on GPU CUDA Cores. For ELL-format sparse tiles, there is a constraint about the total number of the elements (both non-zeros and zeros) kept by

Table 4.7 Transformation primitives

No.	Primitive	Parameter	Description
1.	Reorder rows/columns	–	Reorder the rows/columns of a matrix by #non-zeros per row/column
2.	Divide rows/columns	Size i	Divide the matrix along rows/columns s.t. each part has i rows/columns
3.	Squeeze	–	Only store non-zeros in memory
4.	Split rows	Size j	Split each row s.t. each part has j non-zeros after squeeze
5.	Zero padding	–	Pad zeros along rows/columns
6.	Flatten	–	Store elements into 1D array

a sparse tile like [76], and we only allow sparse tiles satisfying this constraint in the search space.

When doing task partitioning, we split each dense loop (i.e., over which the iteration skips no point) in the computation of a sparse operator evenly into multiple parts, whose sizes are controlled by a parameter, and this determines the task partition way.

The partitioned dense loops and the candidate sparse tiles together determine the candidate sub-tasks. For example, suppose the dense loop of an SpMM operator O, which is loop J, is of length 64 and is partitioned into two parts, each of size 32: $J_1 = \{1, ..., 32\}$, $J_2 = \{33, ..., 64\}$. Let $\{t_1, ..., t_m\}$ be the candidate sparse tiles, where $t_i = (E_i, \alpha_i)$. Then the search space of sub-tasks is $Y = \{x_1, ...x_n\}$, where each $x_i \in Y$ iterates over a subset of the valid points in the iteration space U of the SpMM O, denoted by $u_i \subseteq U$, and u_i is from $\{E_1, ..., E_m\} \times \{J_1, J_2\}$.

Cost Model

We estimate the costs of block-format sub-tasks (sub-tasks based on block-format sparse tiles), the ELL-format sub-tasks (sub-tasks based on ELL-format sparse tiles), and the 1D-format sub-tasks (sub-tasks based on 1D-format sparse tiles) in different ways. For block-format sub-tasks, if the iteration space shapes of two sub-tasks x_1, x_2 are the same, their costs are the same, as they do not skip any point in their iteration space when doing computation (even if they can). Therefore, we list all the iteration space shapes of the block-format sub-tasks our search space contains and measure their costs in advance to build a cost table.

For ELL-format sub-tasks and 1D-format sub-tasks, the cost of a sub-task x is estimated by

Table 4.8 FLOPs, Bytes, and Bytes_{atomic} of a sub-task (#bytes: the unit byte number of the data type (e.g., 2 for float16)). Ind[] maps an element in the ELL format to its original column in the sparse matrix. row[], col[] map an element in the 1D format to its original row and column in the sparse matrix, respectively

Op	Format	Expression	FLOPs	Bytes, $\text{Bytes}_{\text{atomic}}$
SpMM	ELL	C[i,j]=$\sum_k$A[i,k]B[Ind[i,k],j]	2|IJK|	(|IK|+|set(Ind[I,K])|J+|IJ|)×#bytes, Bytes+|IJ|×#bytes
SDDMM	1D	C[i]=$\sum_k$A[row[i],k]B[col[i],k]	2|IK|	(|K|(|set(row[I])|+|set(col[I])|)+|I|)×#bytes, –

$$\text{cost}(x) = g(x) * \text{Atomic}(x).$$

$g(x)$ is the latency to finish x without using the atomic operation (e.g., the "atomicAdd()" function in CUDA) when storing the computation result back to the GPU global memory. The atomic operation is necessary when different GPU threads compute the value of the same address together. For example, given two matrices A, B where A is sparse, suppose A is partitioned into 1×2 blocks, $A = [A_{1,1}, A_{1,2}]$, and B is partition into 2×1 blocks, $B = [B_{1,1}; B_{2,1}]$, then $C = A \times B = A_{1,1} \times B_{1,1} + A_{1,2} \times B_{2,1}$. If we store $A_{1,1}$ in the block format and $A_{1,2}$ in the ELL format, then there must be two sets of GPU threads to compute $A_{1,1} \times B_{1,1}$ and $A_{1,2} \times B_{2,1}$, respectively, and they all write to C, so the write operations need to be atomic. However, it will cause extra overhead because of additional memory operations and the serialization of the conflicting operations. Therefore, we use $\text{Atomic}(x)$ to estimate the influence of atomic operations if any.

Specifically, $g(x) = aR(x) + b$ is a linear function, where a, b are two constants, and $R(x)$ is the roofline model result [74] when not using atomic operations. $R(x)$ is computed by $\frac{\text{FLOPs}}{\min(\text{AI},\text{PEAK})}$: FLOPs is the total floating-point operations performed by x; AI is the arithmetic intensity of x, which is the ratio of FLOPs to the total data movement (Bytes) required to support those FLOPs; PEAK is the ratio of the GPU peak performance (FLOP/s) to the GPU peak bandwidth (Byte/s). Table 4.8 lists how we compute FLOPs and Bytes for the ELL-format sub-tasks and 1D-format sub-tasks we currently support. The values of a, b are obtained by fitting the linear function to the measurement results of different sub-tasks we collect in advance (line 12 of Algorithm 5).

When x does not need atomic operations, $\text{Atomic}(x) = 1$. For 1D-format sub-tasks, $\text{Atomic}(x)$ is usually 1 as they are often used for SDDMM (Sampled Dense-Dense Matrix Multiplication), whose output is sparse and the non-zero output points are often computed independently and in parallel. If there are atomic operations, $\text{Atomic}(x) = cR_{\text{atomic}}(x)/R(x) + d$, where $R_{\text{atomic}}(x)$ is the roofline model result considering atomic operations, which is $\frac{\text{FLOPs}}{\min(\text{FLOPs}/\text{Bytes}_{\text{atomic}},\text{PEAK})}$ and $\text{Bytes}_{\text{atomic}}$ is listed in Table 4.8; c, d are also constant parameters obtained by fitting the function to the pre-collected measurement results (line 13 of Algorithm 5).

Algorithm 5 Get Cost Model Coefficients

Require: None
Ensure: cost model coefficients a, b, c, d

Sample two sets of sub-tasks $\{x_1, ..., x_m\}$, $\{x'_1, ..., x'_m\}$, where x_i, x'_i are the same except that x_i does not do atomic operations while x'_i does
$PEAK \leftarrow \frac{GPU\ peak\ FLOPS}{GPU\ peak\ Byte/s}$
$C, C_{atomic} \leftarrow$ new List, new List
$\mathcal{R}, \mathcal{R}_{atomic} \leftarrow$ new List, new List
for $i = 1, ..., m$ **do**
    $c_1 \leftarrow$ latency of x_i we collect; // details in Section 4.3.3
    $c_2 \leftarrow$ latency of x'_i we collect; // details in Section 4.3.3
    C.append(c_1), C_{atomic}.append(c_2)
    $\mathcal{R}$.append($\frac{FLOPs(x_i)}{\min(FLOPs(x_i)/Bytes(x_i), PEAK)}$)
    $\mathcal{R}_{atomic}$.append($\frac{FLOPs(x'_i)}{\min(FLOPs(x'_i)/Bytes_{atomic}(x'_i), PEAK)}$)
end for
$a, b \leftarrow$ LeastSquaresPolynomialFit($\mathcal{R}, C$)
$c, d \leftarrow$ LeastSquaresPolynomialFit($\mathcal{R}_{atomic}/\mathcal{R}, C_{atomic}/C$)

Greedy with Withdraw

We use a greedy algorithm to solve the MSFD problem. The main idea is as follows. In the beginning, we are given a sparse operator O with its input tensors. We can first get the search space of sub-tasks Y, which is based on the first-level sparse tiles. Let U be the set of points that O cannot skip in its iteration space to get the correct output (Definition 4.3). We set U' to be U initially, which will be updated when selecting sub-tasks. Each sub-task $x_i \in Y$ covers a subset of U, denoted by u_i. We iteratively select sub-tasks to cover all the points in U. At each iteration, we calculate the average cost for x_i by dividing $cost(x_i)$ by $|u_i \cap U'|$ (i.e., the number of the uncovered points of U that x_i can cover). The sub-task with the minimum average cost, x_i, is selected and U' is updated to be $U' - u_i$. As we allow multi-level sparse tiles, after we select a sub-task, we will reorganize the remaining non-zeros of the sparse tensors and generate next-level sparse tiles (using the method in Sect. 4.3.2) and the corresponding new search space of sub-tasks. Then we select the next sub-task from the new sub-tasks. This algorithm stops when there are no uncovered points in U. Based on the above basic workflow, we use some techniques for better effectiveness and efficiency, and Algorithm 6 shows the pseudocode.

Withdraw Sub-tasks (line 8–11, Algorithm 6). The greedy algorithm is sometimes short-sighted. Consider the example in Fig. 4.13. We use rectangles to represent U and the subsets of U covered by sub-tasks. Suppose we already select two sub-tasks x_1, x_2, and the corresponding u_1, u_2 are colored in Fig. 4.13. Suppose the current best sub-task to select is x_3 whose average cost is 1.2. Suppose there is only one SM on the GPU we run. Then we can get a task partition of the original operator with a cost of $8 + 3 + 6 = 17$. However, if we withdraw x_1, x_2 and select x_4 as shown on the lower part of Fig. 4.13, we can get a task partition with a cost of 12,

Algorithm 6 Greedy with Withdraw

Require: operator O, ratio r for local search
Ensure: task partition X

1: $X \leftarrow \emptyset$
2: $Y \leftarrow$ first-level sub-tasks of O
3: $Y' \leftarrow$ subset of Y we may use to replace sub-tasks in X
4: $U \leftarrow$ the set of points O cannot skip in the iteration space
5: $U' \leftarrow U$
6: **while** $|U'| \neq 0$ **do**
7: $\quad x_i \leftarrow \arg\min_{x_j \in Y} \frac{cost(x_j)}{|u_j \cap U'|}, c \leftarrow \min_{x_j \in Y} \frac{cost(x_j)}{|u_j \cap U'|}$ ▷ $Z_j \subseteq X$ is the sub-tasks x_j can replace
8: $\quad c_1 \leftarrow \min_{x_j \in Y'} \frac{cost(x_j) - \sum_{x_k \in Z_j} cost(x_k)}{|u_j \cap U'|}$
9: $\quad$ **if** $c_1 \leq c$ **then**
10: $\quad\quad x_i \leftarrow \arg\min_{x_j \in Y'} \frac{cost(x_j) - \sum_{x_k \in Z_j} cost(x_k)}{|u_j \cap U'|}, c \leftarrow c_1$
11: $\quad$ **end if**
12: $\quad X' \leftarrow$ sub-tasks of Y with average costs $\leq rc$
13: $\quad U' \leftarrow U' - \bigcup_{x_i \in X'} u_i, X \leftarrow X'$
14: $\quad$ Update Y, Y' with the next-level sub-tasks of O
15: **end while**

which is better. Therefore, we adapt the "withdraw" idea from [53] to our greedy algorithm. Specifically, we only consider the block-format non-atomic sub-tasks as the candidates to replace other selected sub-tasks. This is because the example in Fig. 4.13 is due to the unbalanced non-zero distribution of the sparse tensors, and block-format sub-tasks are more likely to encounter this problem as they are obtained without compressing the non-zeros in sparse tensors; on the other hand, we are more interested in using non-atomic sub-tasks to replace atomic sub-tasks because atomic operations are expensive. We compute the average cost of a sub-task x_j to replace a subset of selected sub-tasks Z_j by

$$\frac{\text{cost}(x_j) - \sum_{x_k \in Z_j} \text{cost}(x_k)}{u_j \cap U'}$$

In the example of Fig. 4.13, the average cost of x_4 to replace x_1, x_2 is 0.2, so using this average cost definition, we will select x_4 instead of x_3 in this iteration.

Local Search (line 12, Algorithm 6). To speed up the selection process, every time we select a sub-task x_i, we also search for other sub-tasks whose average costs are close to x_i: less than r times the average cost of x_i, we will select x_j together. Here, r is a user-provided parameter like 120%.

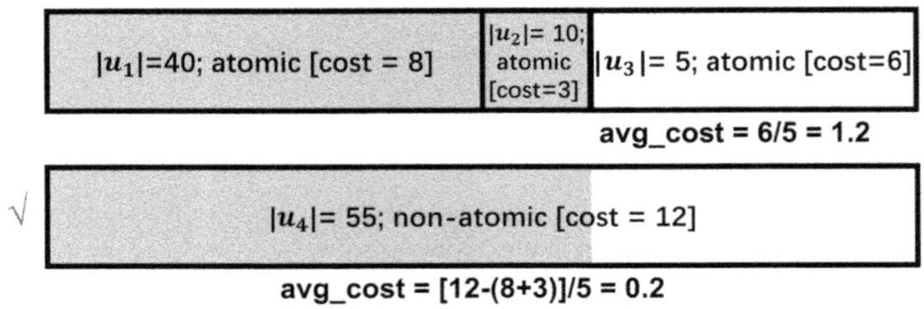

Fig. 4.13 Example to motivate withdrawing sub-tasks

Theoretical Analysis

Theorem 4.2 analyzes the effectiveness of the greedy algorithm for a simplified version of the MSFD problem, where only first-level sparse tiles and the corresponding sub-tasks are considered.

Theorem 4.2 *The approximation ratio of Algorithm 6 for the simplified problem is $r\beta \ln(|U|)$, where U is the set of points that the operator cannot skip in its iteration space to get the correct output, r is the ratio parameter for local search and $\beta < 2$ in most cases.*

Proof The GPU has n SMs in total: $W_1, ..., W_n$. Let X be the finally selected task partition and X^* be the optimal task partition; ALG be $\sum_{x_j \in X} \text{cost}(x_j)$ and OPT be $\sum_{x_j \in X^*} \text{cost}(x_j)$; $u'_j = u_j \cap U'$ be the subset of U that $x_j \in X$ is the first picked sub-task to cover; x_i be $\arg\min_{x_j \in X} \text{cost}(x_j)/|u'_j|$.

We have $\frac{cost(x_i)}{|u'_i|} \leq \frac{ALG}{|U|}$, so $\text{cost}(x_i) \leq \frac{|u'_i|}{|U|} ALG$. Suppose we schedule x_i as the last sub-task to run. The start time of x_i must be no later than $\sum_{x_j \in X - \{x_i\}} \text{cost}(x_j)/n$, as the consequence,

$$cost(X) \leq \sum_{x_j \in X - \{x_i\}} \text{cost}(x_j)/n + \text{cost}(x_i) \leq (\frac{1}{n} + \frac{|u'_i|}{|U|}) ALG \tag{4.5}$$

Let e_k be the kth point in U covered by the greedy algorithm, breaking ties arbitrarily. At the iteration that e_k is covered for the first time, $|U'| \geq |U| - k + 1$. Let X^*_{SUM} be the task partition with the minimum sub-task cost summation, i.e., $\arg\min_{X'} \sum_{x_j \in X'} \text{cost}(x_j)$; OPT_{SUM} be $\sum_{x_j \in X^*_{\text{SUM}}} \text{cost}(x_j)$; u_j be the subset of U that $x_j \in X^*_{\text{SUM}}$ covers. We can evenly assign $\text{cost}(x_j)$ to $u_j \cap U'$ for each $x_j \in X^*_{\text{SUM}}$. Therefore, there must be a $x_l \in X^*_{\text{SUM}}$ s.t. $\frac{\text{cost}(x_l)}{|u_l \cap U'|} \leq \frac{\text{OPT}_{\text{SUM}}}{|U'|}$. Let x_a be the first selected sub-task in X that covers e_k. Let $\text{cost}(e_k)$ be $\frac{\text{cost}(x_a)}{|u'_a|}$ or $\frac{\text{cost}(x_a) - \sum_{x_j \in Z_a} \text{cost}(x_j)}{|u'_a|}$, where Z_a is the sub-tasks replaced by x_a. We have $\text{cost}(e_k)/r \leq \frac{\text{cost}(x_l)}{|u_l \cap U'|} \leq \frac{\text{OPT}_{\text{SUM}}}{|U| - k + 1}$. Therefore, $\text{ALG} = \sum_{k=1}^{|U|} \text{cost}(e_k) \leq \text{OPT}_{\text{SUM}} \cdot r \ln(|U|) \leq \text{OPT} \cdot r \ln(|U|)$. From Eq. 4.5, we have

$$\text{cost}(X) \le (\frac{1}{n} + \frac{|u'_i|}{|U|})r\ln(|U|)\text{OPT} = (1 + \frac{n|u'_i|}{|U|})r\ln(|U|)\frac{\text{OPT}}{n}$$

$$\text{cost}(X) \le (1 + \frac{n|u'_i|}{|U|})r\ln(|U|)\text{cost}(X^*)$$

The subset of U covered by a sub-task in our search space is typically small compared with U in most cases, making $\frac{n|u'_i|}{|U|} < 1$. □

Time Complexity. The number of iterations in Algorithm 6 (line 6–15) is $O(|X|)$, where $|X|$ is the number of the finally selected sub-tasks and $|X| = O(|U|)$. In each iteration, we need to enumerate the candidate sub-tasks and compare their average costs. The number of candidate sub-tasks $|Y| = O(\prod_{i=1}^{d} \ell_i)$, where ℓ_i is the length of the ith dimension of the sparse operator O's iteration space. We also need to update U' in each iteration, which takes $O(|U|)$. The total time complexity is $O(|U|(\prod_{i=1}^{d} \ell_i + |U|))$.

Memory Complexity. We need to store U, U' of the sparse operator O, the currently selected sub-tasks $|X|$, and the candidate sub-tasks in memory, where the u of a sub-task can be obtained from U so we do not need to store it in memory. The total memory complexity is $O(|U| + \prod_{i=1}^{d} \ell_i)$.

4.3.3 Implementation

We prepare templates for block-format, ELL-format, and 1D-format sub-tasks, respectively. The code of different sub-tasks is fused into a single GPU kernel through horizontal fusion [59]. For example, suppose we have two sub-tasks with different code and each sub-task requires a GPU thread block to execute its instructions, then the fused kernel combines the instructions for the two sub-tasks with branch statements. There will be two thread blocks to run the fused kernel, each for one sub-task. The branch condition checks the current thread block ID to dispatch the execution to the path of the instructions of the corresponding sub-task. In the general case, the number of branches in a fused kernel is limited by the number of templates (no larger than 32 in our experiments). These templates are parameterized for some implementation choices, such as the number of threads, the vectorized load/store size, and whether to explicitly unroll the loops. Therefore, we can enumerate the parameters and select the best implementation according to the kernel latency. The parameter space is very small in the experiments (< 100), so the enumeration overhead is low.

For the cost model construction, the cost of a sub-task is measured by running the code based on the template as well. Specifically, we want to collect the running time of a sub-task when the GPU cores are saturated because otherwise estimating the running time of an SM based on summing the costs of sub-tasks up may be less accurate. Therefore, given a sub-task, we create a fake sparse operator whose computation workload can be partitioned into a large enough number of the sub-task and divide the total latency by (#sub-task/#SM) as the cost of the sub-task.

However, if a sparse operator is too sparse that the selected task partition does not have enough sub-tasks, the performance based on the above cost model may not be good enough. Therefore, besides the task partition found by the greedy algorithm, we also search for the best task partition with one format sub-tasks only, e.g., only block-format sub-tasks or only ELL-format sub-tasks or only 1D-format sub-tasks. We compare the latency of these three task partitions and select the best one from them to generate the final code for the sparse operator.

4.3.4 Experiments

The experiments are designed to answer the following questions:

- Can STile outperform the competitors in operator efficiency?
- Are the techniques we use in the greedy algorithm useful?
- Are multi-level decomposition and more than one basic format necessary?

Experiment Setting

Setup. We implement our method based on SparseTIR [76] using Python 3.9 and the code is available on GitHub[1]. The experiments are done on a machine with two AMD EPYC 7413 24-Core Processors, one NVIDIA A100-SXM4-80GB, 1006 GB RAM, Ubuntu 20.04.5, and CUDA 11.7.

Comparison Methods. There are five competitors in our experiments:

- cuSPARSE [47] is an Nvidia library that provides accelerated basic linear algebra subroutines for sparse matrices on GPU. We test the cuSPARSE kernels with the CSR format.
- DGL [71] is a popular framework for GNN. Its SDDMM implementation is based on FeatGraph [56].
- VectorSparse [42] adopts the 1D-vector formats and makes use of GPU tensor cores to accelerate sparse operators.
- Triton [70] is a tiling-based IR for deep learning, and we test its block-sparse operator implementation.
- SparseTIR [76] is a state-of-the-art method that accelerates sparse operators in deep learning with composable formats and composable schedule transformations.
- STile is our method, where r of Algorithm 6 is set to 1.2.

[1] https://github.com/STile-project/STile.

Table 4.9 Tested sparse matrix information

Sparse matrix	M	N	#edges
Cora [69]	2708	2708	10,556
Citeseer [69]	3327	3327	9228
Pumbed [69]	19,717	19,717	88,651
ppi [51]	44,906	44,906	1,271,274
arxiv [67]	169,343	169,343	1,166,243
NotreDame (ND) [39]	325,729	325,729	1,497,134
Proteins [67]	132,534	132,534	39,561,252
Reddit [51]	232,965	232,965	114,615,892
Unstructured	768–3072	768–3072	2950–418,402
Structured	128–3072	128–3072	6144–503,808
LogSparse [60]	4096	4096	90,115
Strided [43]	4096	4096	766,144

The first four competitors only support single-sparse-format operators, while SparseTIR supports hybrid formats. As done in the experiments of SparseTIR, we test Triton only on the block-pruned matrices and the sparse attention matrices. SparseTIR uses different sparse formats for different sparse matrices, for which we also refer to its experiment in our comparison.

Datasets. We test the performance of our method on two kinds of operators: Sparse Matrix Multiplication (SpMM) and Sampled Dense-Dense Matrix Multiplication (SDDMM). Table 4.9 lists the information of the sparse matrices used in the experiments, which can be divided into four categories: (1) the graph adjacency matrices used for Graph Neural Networks with different sizes, densities and sparse patterns, i.e., the first eight sparse matrices (from cora to reddit) in Table 4.9; (2) the sparse matrices obtained by performing unstructured pruning to the weight matrices of Transformers, thus having unstructured sparse patterns, denoted by "unstructured" in Table 4.9; (3) the sparse matrices obtained by performing structured pruning to the weight matrices of Transformers, thus having structured sparse patterns (block-sparse patterns, specifically), denoted by "structured" in Table 4.9; (4) the sparse attention matrices used for Transformers with different sparse patterns, i.e., LogSparse, Strided in Table 4.9. There are 72 unstructured sparse matrices extracted from the movement-pruned PruneBERT[2] and 72 structured sparse matrices extracted from the block-pruned PruneBERT[3] with block size 32. The data type in the experiments is float16.

[2] https://huggingface.co/huggingface/prunebert-base-uncased-6-finepruned-w-distil-squad.

[3] https://huggingface.co/madlag/bert-base-uncased-squad1.1-block-sparse-0.07-v1.

Main Results

This subsection compares our method and the competitors on SpMM and SDDMM in terms of the operator latency and the operator memory usage. We also report the search time of Algorithm 6.

SpMM

We select cuSPARSE, VectorSparse, Triton, SparseTIR as the competitors, where Triton is only used for the structured sparse matrices, LogSparse and Strided. Figs. 4.14 and 4.15 show the geometric mean speedups and the normalized memory usage values of different methods against cuSPARSE, respectively.

Graph Adjacency Matrix. For each graph adjacency matrix, the feature size of the SpMM operator is set to J, where $J \in \{32, 64, 128, 256, 512\}$. Our method achieves the best operator latency on each sparse matrix. We are $2.1 - 18.0\times$ better than cuSPARSE, $2.3 - 5.3\times$ better than VectorSparse, $1.4 - 2.7\times$ better than SparseTIR. SparseTIR uses the hyb format for the sparse matrices. The hyb format partitions the sparse matrix into several parts along columns first, then puts the rows into different buckets according to the #non-zeros per row. Each bucket is stored in a separate ELL format. The hyb format can effectively improve the cache locality and the load balance, avoiding having too much computation waste due to the padding in ELL at the same time. However, SparseTIR with hyb format does not use GPU Tensor Cores. VectorSparse adopts the 1D-vector format, where a vector is a column of elements with specific height (including both zeros and non-zeros in the sparse matrix). VectorSparse stores a sparse matrix into multiple vectors and uses Tensor Cores to do matrix multiplication. Tensor Cores are special computation units that are much more efficient than the GPU CUDA cores, but they only accept dense matrices of specific shapes as the input for matrix multiplication. This limits their application to sparse matrix operations. Although the 1D-vector format needs to store fewer zeros (i.e., less computation waste) than simply using the block-format for Tensor Core computation, it is still not good enough when the sparse matrix is not dense enough.

Compared with these competitors, STile has higher speedups for several reasons. First, we consider both the block-format sub-tasks and the ELL-format sub-tasks in our search space and try to find the best combination of them. In this way, we can use Tensor Cores only for the dense part of the sparse matrix, i.e., the block-format sub-tasks. Second, our templates for the ELL-format sub-tasks make use of the GPU shared memory to reduce the expensive data movement from the GPU global memory to the local registers. Third, we can reorder the matrix rows and columns to improve the local density of the sparse matrix, and our search space includes more possible block shapes for the block-format sub-tasks, which enables us to find better hybrid formats.

In terms of the operator memory usage, compared with the CSR format adopted by cuSPARSE, "hyb" and "ELL" do not need the row length array as their row lengths are constants; "1D-vector" and "block" store the positions for vectors or blocks (rather than for non-zero elements) in the CSR way, which cost less memory. However, CSR does not pad zeros, unlike all the other formats. This explains why VectorSparse, SparseTIR and STile can save memory on some datasets (e.g., cora, citeseer) compared with cuSPARSE, while not on other datasets (e.g., proteins, reddit). STile costs much more memory on ppi, proteins and reddit than all the baselines, because on these datasets, the majority of the selected sub-tasks adopt the block format. By contrast, the length of each 1D-vector used by VectorSparse is shorter than our block height, so it is more flexible than the blocks we use; the hyb format of SparseTIR is also more memory efficient as it requires fewer zero paddings. Although we store more zeros than other methods on these datasets, thanks to the Tensor Core computation enabled by the block format and our efficient kernel templates, we still achieve lower operator latency.

Unstructured Sparse Matrix. We extract 72 SpMM operators from the movement-pruned PruneBERT, with the sequence length set to 512. SparseTIR uses the BSR format or the SR-BCRS format [61] (choosing the better one from them). The BSR format stores the blocks of a sparse matrix with at least one non-zero. The SR-BCRS format is similar to the 1D-vector sparse format of VectorSparse, but it stores the vectors in a stride-wise row-major manner, while VectorSparse stores the vectors in a consecutive array. Both the BSR format and the SR-BCRS format of SparseTIR perform computation on Tensor Cores. Figure 4.14 shows that, on average, we are 3.1× better than cuSPAESE, 15.5× better than VectorSparse, and 1.3× better than SparseTIR. Although VectorSparse, SparseTIR, and our method all support Tensor Core computation in this part of the experiments, our operator latency is still better than theirs because we allow more flexible matrix reordering and more possible block-format sub-tasks, i.e., more possible computation schedules. In terms of operator memory usage, VectorSparse outperforms STile for the reason mentioned in the experiments on graph adjacency matrices. However, our block height is shorter than the BSR block height and the SR-BCRS vector length in SparseTIR, so our blocks are more flexible, which together with the flexible matrix reordering makes our format use less memory than SparseTIR.

Structured Sparse Matrix. We extract 72 SpMM operators from the block-pruned PruneBERT (sequence length: 512). SparseTIR here uses the BSR format or the DBSR (doubly-compressed BSR) format for a sparse matrix. The DBSR format is similar to the BSR format but skips the all-zero rows. Triton is also included as the competitor in this part of the experiment, with the block size set to 32, which is exactly the block size used in the block pruning. Figure 4.14 shows that, on average, we are 7.6× better than cuSPARSE, 7.6× better than VectorSparse, 1.3× better than Triton, and 1.1× better than SparseTIR. The competitors, VectorSparse, Triton, and SparseTIR, all use Tensor Cores for computation. The reason that we perform better than them is similar to that for the experiments of unstructured sparse matrices. The

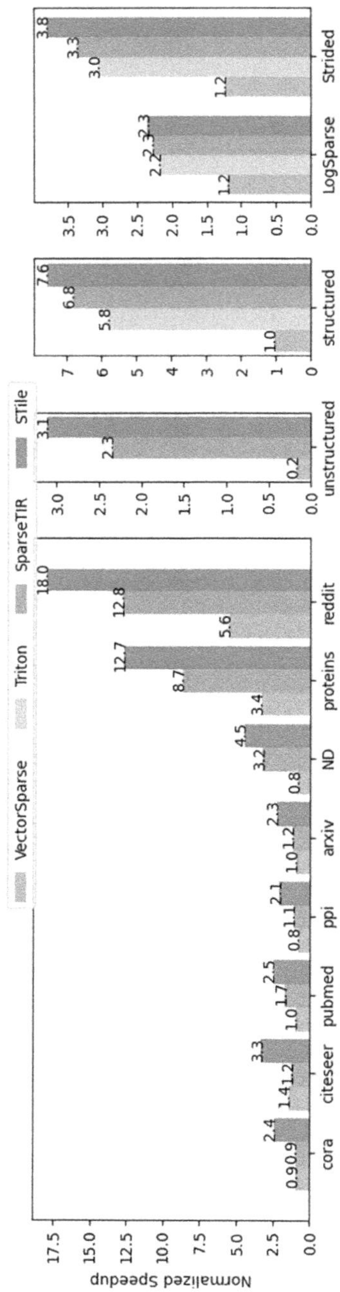

Fig. 4.14 Normalized speedup against cuSPARSE for SPMM

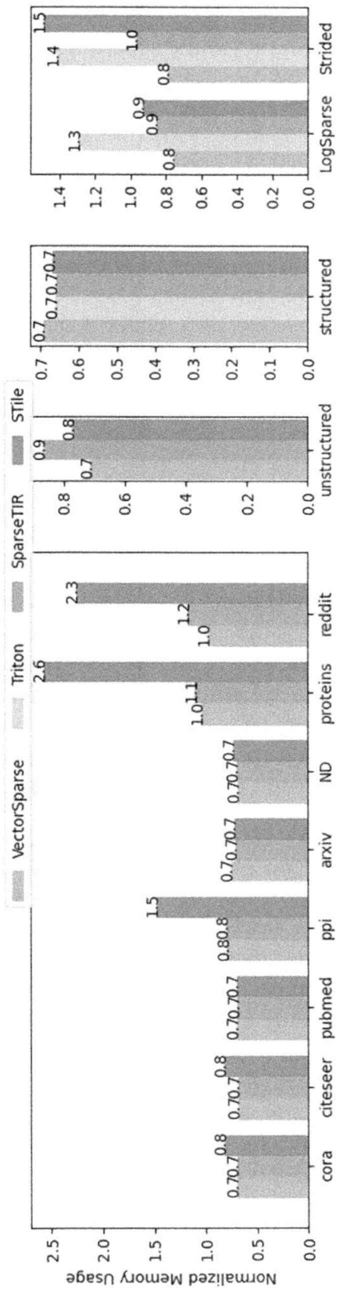

Fig. 4.15 Normalized memory usage against cuSPARSE for SPMM

operator memory usage values of all methods except cuSPARSE are close, because their formats (storing non-zero vectors or blocks) suit the block-sparse matrices well.

Sparse Attention Matrix. In this part of the experiment, the SpMM operator feature size is set to J, where $J \in \{32, 64, 128, 256, 512\}$. The sparse patterns of LogSparse and Strided consist of sliding windows and diagonals. Triton chooses its block size from 16 and 32. SparseTIR selects the best format from CSR, BSR, DBSR, and SR-BCRS, and the formats selected are BSR (for one LogSparse SpMM) and SR-BCRS (for all other SpMMs). For Strided SpMMs, STile only selects block-format sub-tasks; for LogSparse SpMMs, our formats either consist of block basic formats only, or ELL basic formats only. Figure 4.14 shows that the operator latency we achive is $2.3 - 3.8\times$ better than cuSPARSE, $2.0 - 3.1\times$ better than VectorSparse, $1.1 - 1.2\times$ better than Triton, $1.0 - 1.1\times$ better than SparseTIR. On Strided, our format costs more memory than the SR-BCRS of SparseTIR, despite our shorter block height because we select blocks with large widths for better operator latency, which requires more padding.

SDDMM

We now compare the operator latency and operator memory usage of our method and the competitors on SDDMM operators. We select DGL, VectorSpaese, SparseTIR (BSR), SparseTIR, and Triton as the competitors. SparseTIR (BSR) uses the BSR format for sparse matrices, while SparseTIR uses the 1D format for sparse matrices, which stores the non-zeros in a 1D array. SparseTIR (BSR) is not used for graph adjacency matrices. Triton is used for structured sparse matrices, LogSparse and Strided. Figs. 4.16 and 4.17 show the geometric mean speedups and the normalized operator memory usage of different methods against DGL, respectively.

Graph Adjacency Matrix. For each graph adjacency matrix, the feature size of the SDDMM operator is set to K, where $K \in \{32, 64, 128, 256, 512\}$. Figure 4.16 shows that our method outperforms all the competitors on all the sparse matrices. Specifically, we are $1.5 - 6.9\times$ better than DGL, $3.4 - 2283.5\times$ better than VectorSparse, and $1.1 - 1.3\times$ better than SparseTIR. For SDDMM, the computation workloads of different output points are independent of each other, and the workload amounts are the same. Thus, using the 1D format for a sparse matrix and assigning the computation of different output points to threads does not have the problem of workload balance. The optimization techniques like vectorized load/store intrinsics and parallelization on the reduction loop also help SparseTIR perform better. Therefore, SparseTIR achieves $1.4 - 5.6\times$ speedup against DGL. VectorSparse runs the computation on Tensor Cores, and it also implements the reduction-loop parallelization, but its performance is much worse, which shows simply using the 1D-vector format is not a good choice for these tested matrices. Our method considers both block-format sub-tasks (running on Tensor Cores) and 1D-format sub-tasks in the search space, which helps us take advantage of both and perform better than the competitors. For example, on reddit with $K = 32$, if we only consider 1D-format sub-tasks, the SDDMM latency is 3.24 ms; if we only consider block-format sub-tasks, the latency is 2.66 ms; when considering both, the latency is 2.07 ms, i.e., 1.6× and 1.3× better than the first two cases, respectively. For the operator memory usage, the 1D format stores the row and column indices for each non-zero element, while the block format or the 1D-vector format uses less memory for positions of blocks or vectors but may store extra zeros. In Fig. 4.17, SparseTIR performs the same as DGL because they both use the 1D format. VectorSparse performs worse ($1.0 - 1.3\times$) because it selects a large vector length for better operator latency, causing a lot of padding. STile performs the same or better than DGL (0.9× on reddit) because our hybrid format saves memory for position information on denser parts and avoids padding on sparser parts.

Unstructured Sparse Matrix. The SDDMM operators take the unstructured sparse matrices as input, and the reduction-loop length is set to $K = 512$. Figure 4.16 shows that we are 2.4× better than DGL, 4.8× better than VectorSparse, 3.4× better than SparseTIR (BSR), and 1.1× better than SparseTIR. The hybrid formats finally selected by our method only include the 1D-format sparse tiles. However, we still outperform SparseTIR because the parameter space of our templates for the 1D-format sub-tasks is larger than SparseTIR. Although STile, SparseTIR and DGL all use the 1D format, STile and SparseTIR lead to less operator memory because DGL pads the two input matrices for multiplication to be of the same shape.

Structured Sparse Matrix. The SDDMM operators take the structured sparse matrices as input, and the reduction-loop length is set to $K = 512$. Figure 4.16 shows that the operator latency we achieve is 4.3× better than DGL, 3.6× better than VectorSparse, 1.7× better than SparseTIR (BSR), 1.6× better than SparseTIR, and 1.1× better than Triton. We are better than SparseTIR (with 1D format) because we support Tensor Cores, which are suitable for block-sparse patterns. We are better than other competitors that also run on Tensor Cores because our candidate block-format sub-tasks in the search space implemented with our templates require fewer resources (e.g., threads) and are more efficient. The operator memory usage values of all the methods are only $0.5 - 0.7\times$ of DGL.

Sparse Attention Matrix. The SDDMM operators take the sparse attention matrices as input, and the feature size is set to K, where $K \in \{32, 64, 128, 256, 512\}$. Figure 4.16 shows that the operator latency we achieve is $3.3 - 7.5\times$ better than DGL, $3.5 - 5.2\times$ better than VectorSparse, $1.7 - 1.9\times$ better than SparseTIR (BSR), $1.2 - 1.7\times$ better than SparseTIR, and $1.5 - 1.7\times$ better than Triton. We only select 1D-format spare tiles for LogSparse. For Strided, we select both block-format and 1D-format sparse tiles, which also makes our format the most memory-efficient among the competitors.

Search Time

Table 4.10 shows the average search time of Algorithm 6 for SPMM and SDDMM operators. The search time is consistent with our time complexity analysis

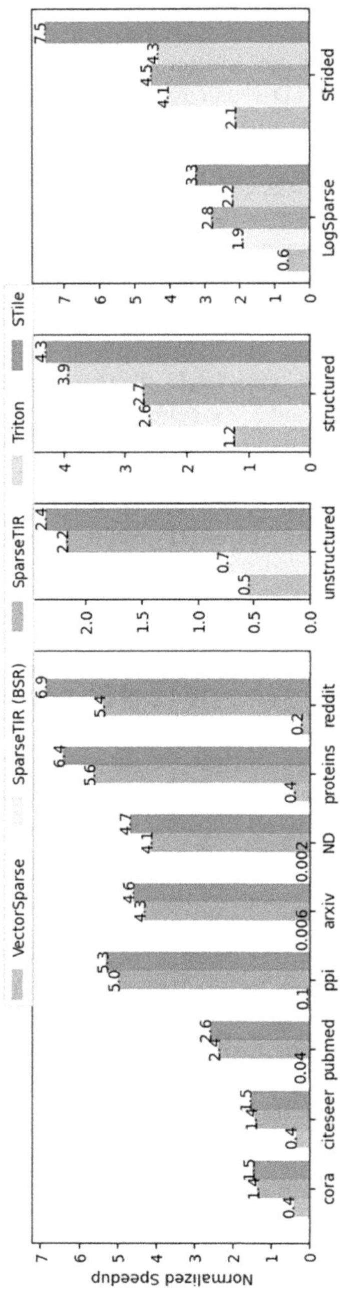

Fig. 4.16 Normalized speedup against DGL for SDDMM

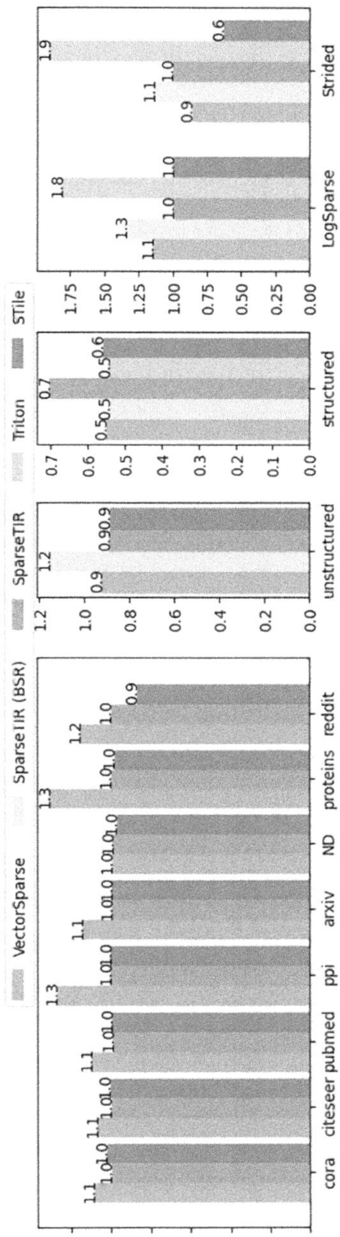

Fig. 4.17 Normalized memory usage against DGL for SDDMM

Table 4.10 The search time (seconds) of Algorithm 6

Sparse matrix	SpMM	SDDMM
Cora	0.5	0.3
Citeseer	0.6	0.6
Pumbed	93.9	4.9
ppi	88.0	28.2
arxiv	1154.7	247.9
ND	3015.7	732.8
Proteins	1161.8	1242.6
Reddit	2831.0	1772.5
Unstructured	0.1–12.9	0.1–3.4
Structured	0.03–0.4	0.03–0.9
LogSparse	1.0	0.7
Strided	2.3	1.7

in Sect. 4.3.2. We can finish the search for any of the tested datasets in one hour. For those small sparse matrices (e.g., citeseer, cora), the search time is less than one second. Our search time is acceptable as it can be amortized when the optimized sparse operators are executed multiple times during the inference or training tasks. Other competitors do not need tuning or only tune in a very small parameter space, so we do not report their search time results.

Ablation Study

This section studies the influence of the optimization techniques we propose for Algorithm 6: withdrawing sub-tasks and local search.

With Withdrawing versus Without Withdrawing. We compare the performance of our greedy algorithm with and without the withdrawing step for SpMMs on the graph adjacency matrices. The feature size is set to $J = 32$. The second column of Table 4.11 shows the normalized latency against the "with withdrawing" setting. Without withdrawing, the SpMM latency is $1.0 - 1.5\times$ higher, which verifies that the withdrawing step can effectively improve the performance of the greedy algorithm. When analyzing the finally selected sub-tasks in different settings, we find that when without withdrawing, the greedy algorithm tends to select more atomic sub-tasks, which makes the latency higher. For example, on proteins, without withdrawing, there are 81745 atomic block-format sub-tasks selected by the greedy algorithm; while with withdrawing, only one atomic block-format sub-task is selected by the greedy algorithm.

With Local Search versus Without Local Search. The local search ratio r in Algorithm 6 is set to 1.2 by default. In this part of experiments, we compare the perfor-

Table 4.11 Normalized latency and search time against the default setting of STile

Sparse matrix	w/o withdrawing	w/o local search	
	Latency	Latency	Search time
Cora	1.0	1.0	1.0
Citeseer	1.0	1.0	1.3
Pumbed	1.0	1.0	2.2
ppi	1.4	1.2	61.7
arxiv	1.1	1.0	22.1
ND	1.2	1.2	56.1
Proteins	1.5	1.0	213.5
Reddit	1.1	1.0	267.4

Table 4.12 Normalized latency and search time against the default setting of STile

Sparse	Operator	Latency						Search time					
Matrix		Level num				Basic format		Level num				Basic format	
Reddit	SpMM	1.1	1.0	1.0	1.0	1.1	1.9	1.2	0.8	0.9	1.0	0.2	0.1
	SDDMM	1.4	1.3	1.0	1.0	2.3	1.1	2.0	1.8	1.0	1.0	0.3	0.1
ND	SpMM	1.7	1.4	1.6	1.0	1.1	1.3	1.7	2.7	1.4	1.0	0.1	0.01
	SDDMM	1.1	1.1	1.0	1.0	1.6	1.1	2.7	2.1	1.0	1.0	0.1	0.002
LogSparse	SpMM	1.0	1.0	1.0	1.0	1.0	1.0	0.9	1.0	1.0	1.0	0.7	0.1
	SDDMM	1.3	1.0	1.0	1.0	2.1	1.0	1.3	1.1	1.1	1.1	0.2	0.1
Strided	SpMM	1.0	1.0	1.0	1.0	1.0	2.0	1.0	1.0	1.0	1.0	0.7	0.3
	SDDMM	1.2	1.0	1.0	1.0	2.6	1.4	2.5	1.1	1.1	1.0	0.2	0.4

mance of Algorithm 6 when with and without local search on the SpMM operators taking the graph adjacency matrices as input, and report the results in Table 4.11 (the last two columns). When not doing the local search, the SpMM latency is the same or even higher, e.g., $1.2\times$ of the latency when doing the local search. By looking at the sub-tasks selected for ppi, we find that Algorithm 6 selects more atomic sub-tasks when not doing the local search, so the latency is higher. Regarding the search time, doing local search greatly speeds up the search process, i.e., $1.0 - 267.4\times$ speedup. Local search enables us to tackle the optimization of larger sparse operators.

Search Space Study

This section shows how the search space depth and breadth influence the final operator latency and the search time of Algorithm 6. Specifically, the search space depth is controlled by the number of decomposition levels, which is set to values in $\{1, 2, 10, 100, \infty\}$, where ∞ is adopted by STile by default. The search space breadth is controlled by the set of basic sparse formats we consider, which is varied from one basic format to two basic formats (the default setting). Table 4.12 shows the results on two datasets with natural hybrid sparse patterns (reddit, ND) and two datasets with manually designed hybrid sparse patterns (LogSparse, Strided). The feature size is set to 64 for SpMM and 256 for SDDMM.

For both SpMM and SDDMM, the operator latency generally improves as the depth increases and becomes stable after the depth is large enough (the default setting is $1.0 - 1.7\times$ better than the single-level setting). For SpMM on ND, the operator latency when the depth is 10 is longer than that when the depth is 2 because more atomic sub-tasks are selected when the depth is 10, which is possible because Algorithm 6 is not guaranteed to find the optimal task partition. The best depth for SpMM on LogSparse and Strided is 1 (i.e., all sparse tiles are extracted from the original sparse matrix), so changing depth does not influence the operator latency. The best depth is determined by the number of distinct sparse patterns in an operator, which varies from 1 to 24 for the tested operators in Sect. 4.3.4. With the depth set to ∞, our algorithm will adaptively explore the depth according to the cost model. However, although increasing depth extends the search space, the search time of Algorithm 6 does not necessarily increase, e.g., SDDMM on Strided. The reason is that although setting a larger depth increases the time of generating next-level sparse tiles, the remaining non-zeros will be reorganized in this process, making the new sparse tiles more effective in covering these non-zeros, which speeds up the search process, as some old sparse tiles may become partially useful if the non-zeros they cover have already been covered by the selected sparse tiles. For example, for SDDMM on Strided, the search process under the single-level setting takes 15 iterations and selects 13720 sparse tiles in total, while the search process of the multi-level setting takes 2 iterations only and selects 9848 sparse tiles in total, which explains the search time difference.

Increasing the search space breadth improves the operator latency for both SpMM and SDDMM in general (considering two basic formats is $1.0 - 2.6\times$ better than considering one basic format). Therefore, we appreciate the set of basic sparse formats to be efficient for different sparse patterns, e.g., the block format and ELL for SpMM work well on high density and low density, respectively, so their combinations work well on various sparse patterns. On the other hand, setting the search space breadth to 1 significantly reduces the search space size, so the search time of Algorithm 6 is $0.002 - 0.7\times$ of the default setting for SDDMM on ND.

Figure 4.18 shows the sparse formats for SDDMM on Strided that we can obtain in different settings (we use a small matrix with the Strided sparse pattern for illustration). The green elements are stored in the block format, while the blue ones are in the 1D format. In the default setting, we can separate the dense part and the sparse part of the sparse matrix A well into two parts and store them in suitable formats respectively. When the depth is 1, to store the non-zeros in the sparser part, the format has to redundantly store some non-zeros as we cannot generate new 1D-format sparse tiles for the remaining non-zeros, which wastes computation. Only considering the block basic format also wastes computation; while only considering the 1D basic format fails to use the Tensor Cores.

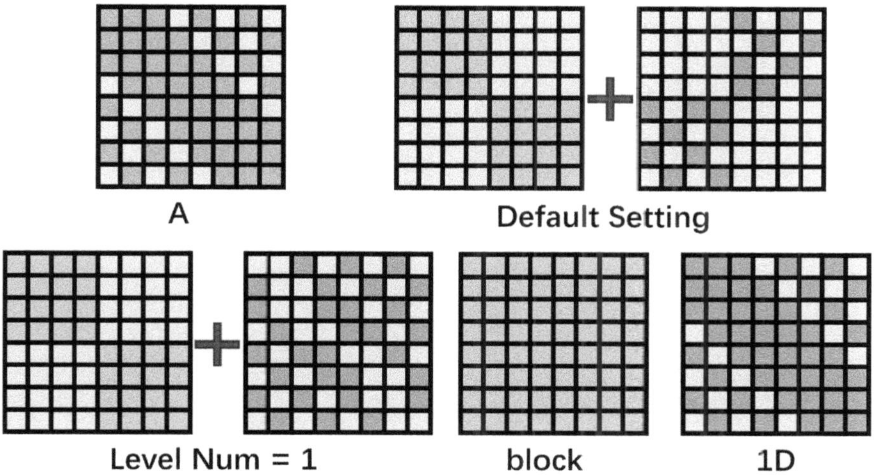

Fig. 4.18 Sparse formats obtained in different settings

Cost Model Performance

To show that our cost model is effective, we randomly sample 5000 sub-tasks for SpMM and 13060 sub-tasks for SDDMM from the sub-tasks we select. The sampled sub-tasks are all ELL-format for SpMM and 1D-format for SDDMM, because for block-format sub-task, we already built a cost table to look for their costs. We measure the latency of each sub-task on GPU when all the GPU cores are saturated using the method in Sect. 4.3.3. Figure 4.19 compares the real latency values and the predicted costs of our cost model for the ELL-format sub-tasks and for the 1D-format sub-tasks, respectively. There is a linear correlation between our prediction and the real value. The Pearson correlation coefficients of the cost model for the ELL-format sub-tasks and for the 1D-format sub-tasks are 0.9243 and 0.9997, respectively, and the p-values are 0.0 for both. The RMSE results are 1.4e-4 and 4.8e-05 for these two types of sub-tasks, respectively. The cost model performs very well on 1D-format sub-tasks because we only consider one type of 1D-format sparse tiles in our search space. On the other hand, although the cost model for ELL-format sub-tasks is less accurate (the empirical approximation bound is 24%), given that block-format sub-tasks with Tensor Core computation can be a magnitude faster, this cost model still helps us find good hybrid formats for SpMMs. We also try MLP [64] as the cost model for ELL-format sub-tasks. The RMSE of MLP is 3.3e-05, which is almost a magnitude smaller than ours. The operator latency values we achieve on the datasets in Sect. 4.3.4 are the same as those when using our cost model, but the search time increases a lot ($1.1 - 6.1\times$) because we need to compute more features for MLP to do prediction. Therefore, our cost model achieves good performance in predicting operator latency while remaining efficient to execute.

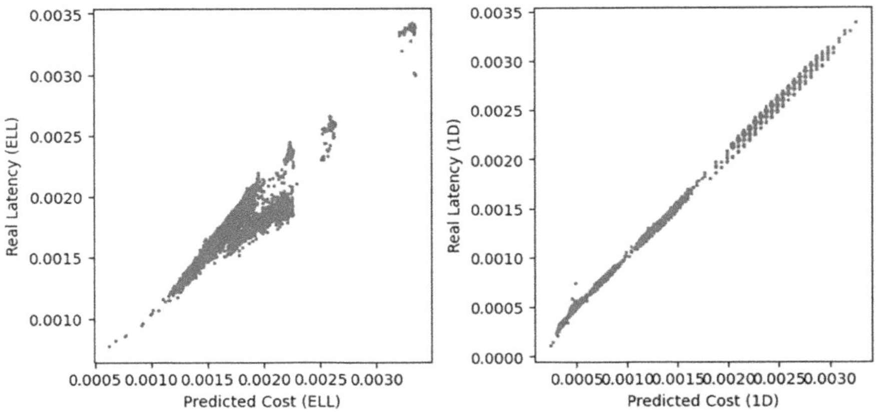

Fig. 4.19 1D-format sub-task: real latency versus predicted cost

4.4 Conclusion

This chapter presented two methods for accelerating computation for scalable GNN training: (1) DAHA, a model-level framework for fine-grained, data- and hardware-aware execution planning and (2) STile, an operator-level framework for automatically optimizing sparse computations. DAHA introduces a novel hybrid CPU-GPU pipeline, intra-batch execution plan rewriting, and efficient inter-batch scheduling, achieving significant end-to-end speedups. Meanwhile, STile addresses the hybrid sparse format selection problem by exploring an expanded search space via flexible tensor transformations and multi-level tiling, yielding up to 18x speedup over cuSPARSE for SpMM. Future work includes integrating caching strategies into DAHA and further optimizing the search process.

References

1. Sergi Abadal, Akshay Jain, Robert Guirado, Jorge López-Alonso, and Eduard Alarcón. 2021. Computing graph neural networks: A survey from algorithms to accelerators. *ACM Computing Surveys (CSUR)* (2021).
2. David Applegate and William Cook. 1991. A computational study of the job-shop scheduling problem. *ORSA Journal on computing* 3, 2 (1991), 149–156.
3. Bo Chen, Chris N Potts, and Gerhard J Woeginger. 1998. A review of machine scheduling: Complexity, algorithms and approximability. *Handbook of Combinatorial Optimization: Volume1–3* (1998), 1493–1641.
4. Jie Chen, Tengfei Ma, and Cao Xiao. 2018a. FastGCN: Fast Learning with Graph Convolutional Networks via Importance Sampling. In *ICLR*.
5. Jianfei Chen, Jun Zhu, and Le Song. 2018b. Stochastic Training of Graph Convolutional Networks with Variance Reduction. In *ICML*.

6. Wei-Lin Chiang, Xuanqing Liu, Si Si, Yang Li, Samy Bengio, and Cho-Jui Hsieh. 2019. Cluster-gcn: An efficient algorithm for training deep and large graph convolutional networks. In *SIGKDD*.
7. Weilin Cong, Rana Forsati, Mahmut Kandemir, and Mehrdad Mahdavi. 2020. Minimal Variance Sampling with Provable Guarantees for Fast Training of Graph Neural Networks. In *SIGKDD*.
8. Matthias Fey and Jan Eric Lenssen. 2019. Fast Graph Representation Learning with PyTorch Geometric. http://arxiv.org/abs/1903.02428
9. Aleksei V Fishkin, Klaus Jansen, and Monaldo Mastrolilli. 2008. Grouping techniques for scheduling problems: Simpler and faster. *Algorithmica* 51, 2 (2008), 183–199.
10. Swapnil Gandhi and Anand Padmanabha Iyer. 2021. P3: Distributed Deep Graph Learning at Scale. In *OSDI*.
11. Leslie Ann Goldberg, Mike Paterson, Aravind Srinivasan, and Elizabeth Sweedyk. 2001. Better approximation guarantees for job-shop scheduling. *SIAM Journal on Discrete Mathematics* 14, 1 (2001), 67–92.
12. Will Hamilton, Zhitao Ying, and Jure Leskovec. 2017. Inductive representation learning on large graphs. In *NIPS*.
13. Ellis Horowitz and Sartaj Sahni. 1976. Exact and Approximate Algorithms for Scheduling Nonidentical Processors. *J. ACM* 23, 2 (apr 1976), 317–327. https://doi.org/10.1145/321941.321951
14. Weihua Hu, Matthias Fey, Marinka Zitnik, Yuxiao Dong, Hongyu Ren, Bowen Liu, Michele Catasta, and Jure Leskovec. 2020. Open graph benchmark: Datasets for machine learning on graphs. In *NIPS*.
15. Wenbing Huang, Tong Zhang, Yu Rong, and Junzhou Huang. 2018. Adaptive sampling towards fast graph representation learning. In *NIPS*.
16. Tim Kaler, Nickolas Stathas, Anne Ouyang, Alexandros-Stavros Iliopoulos, Tao Schardl, Charles E Leiserson, and Jie Chen. 2022. Accelerating training and inference of graph neural networks with fast sampling and pipelining. In *MLSys*.
17. Qifa Ke, Michael Isard, and Yuan Yu. 2013. Optimus: a dynamic rewriting framework for data-parallel execution plans. In *EuroSys*.
18. Taehyun Kim, Changho Hwang, KyoungSoo Park, Zhiqi Lin, Peng Cheng, Youshan Miao, Lingxiao Ma, and Yongqiang Xiong. 2021. Accelerating gnn training with locality-aware partial execution. In *Proceedings of the 12th ACM SIGOPS Asia-Pacific Workshop on Systems*. 34–41.
19. Thomas N. Kipf and Max Welling. 2017. Semi-Supervised Classification with Graph Convolutional Networks. In *ICLR*.
20. Zhiqi Lin, Cheng Li, Youshan Miao, Yunxin Liu, and Yinlong Xu. 2020. Pagraph: Scaling gnn training on large graphs via computation-aware caching. In *SoCC*.
21. Zekun Lu, Qiancheng Yu, Xia Li, Xiaoning Li, and Qinwen Yang. 2023. Learning Weight Signed Network Embedding with Graph Neural Networks. *Data Science and Engineering* 8, 1 (2023), 36–46.
22. Adam Paszke, Sam Gross, Francisco Massa, Adam Lerer, James Bradbury, Gregory Chanan, Trevor Killeen, Zeming Lin, Natalia Gimelshein, Luca Antiga, Alban Desmaison, Andreas Kopf, Edward Yang, Zachary DeVito, Martin Raison, Alykhan Tejani, Sasank Chilamkurthy, Benoit Steiner, Lu Fang, Junjie Bai, and Soumith Chintala. 2019. PyTorch: An Imperative Style, High-Performance Deep Learning Library. In *NIPS*.
23. Oleksandr Shchur, Maximilian Mumme, Aleksandar Bojchevski, and Stephan Günnemann. 2018. Pitfalls of graph neural network evaluation. *arXiv preprint* arXiv:1811.05868 (2018).
24. John Thorpe, Yifan Qiao, Jonathan Eyolfson, Shen Teng, Guanzhou Hu, Zhihao Jia, Jinliang Wei, Keval Vora, Ravi Netravali, Miryung Kim, and Guoqing Harry Xu. 2021. Dorylus: Affordable, Scalable, and Accurate GNN Training with Distributed CPU Servers and Serverless Threads. In *OSDI*.
25. Xinchen Wan, Kaiqiang Xu, Xudong Liao, Yilun Jin, Kai Chen, and Xin Jin. 2023. Scalable and Efficient Full-Graph GNN Training for Large Graphs. In *SIGMOD*.

26. Minjie Wang, Da Zheng, Zihao Ye, Quan Gan, Mufei Li, Xiang Song, Jinjing Zhou, Chao Ma, Lingfan Yu, Yu Gai, Tianjun Xiao, Tong He, George Karypis, Jinyang Li, and Zheng Zhang. 2019. Deep Graph Library: A Graph-Centric, Highly-Performant Package for Graph Neural Networks. *arXiv preprint* arXiv:1909.01315 (2019).
27. Jingshu Peng, Zhao Chen, Yingxia Shao, Yanyan Shen, Lei Chen, and Jiannong Cao. 2022. SANCUS: Staleness-Aware Communication-Avoiding Full-Graph Decentralized Training in Large-Scale Graph Neural Networks. *VLDB* 15, 9 (2022), 1937–1950. https://doi.org/10.14778/3538598.3538614
28. Qiange Wang, Yanfeng Zhang, Hao Wang, Chaoyi Chen, Xiaodong Zhang, and Ge Yu. 2022. NeutronStar: Distributed GNN Training with Hybrid Dependency Management. In *SIGMOD*.
29. Yongji Wu, Matthew Lentz, Danyang Zhuo, and Yao Lu. 2022. Serving and Optimizing Machine Learning Workflows on Heterogeneous Infrastructures. In *PVLDB*.
30. Shuo Xiao, Dongqing Zhu, Chaogang Tang, and Zhenzhen Huang. 2023. Combining Graph Contrastive Embedding and Multi-head Cross-Attention Transfer for Cross-Domain Recommendation. *Data Science and Engineering* 8, 3 (2023), 247–262.
31. Jianbang Yang, Dahai Tang, Xiaoniu Song, Lei Wang, Qiang Yin, Rong Chen, Wenyuan Yu, and Jingren Zhou. 2022. GNNLab: A Factored System for Sample-Based GNN Training over GPUs. In *EuroSys*.
32. Rex Ying, Ruining He, Kaifeng Chen, Pong Eksombatchai, William L Hamilton, and Jure Leskovec. 2018. Graph convolutional neural networks for web-scale recommender systems. In *SIGKDD*.
33. Hanqing Zeng, Hongkuan Zhou, Ajitesh Srivastava, Rajgopal Kannan, and Viktor Prasanna. 2020. GraphSAINT: Graph Sampling Based Inductive Learning Method. In *ICLR*.
34. Dalong Zhang, Xin Huang, Ziqi Liu, Jun Zhou, Zhiyang Hu, Xianzheng Song, Zhibang Ge, Lin Wang, Zhiqiang Zhang, and Yuan Qi. 2020. AGL: A Scalable System for Industrial-Purpose Graph Machine Learning. In *PVLDB*.
35. Chenguang Zheng, Hongzhi Chen, Yuxuan Cheng, Zhezheng Song, Yifan Wu, Changji Li, James Cheng, Hao Yang, and Shuai Zhang. 2022a. ByteGNN: efficient graph neural network training at large scale. In *PVLDB*.
36. Da Zheng, Xiang Song, Chengru Yang, Dominique LaSalle, and George Karypis. 2022b. Distributed hybrid CPU and GPU training for graph neural networks on billion-scale heterogeneous graphs. In *SIGKDD*.
37. Rong Zhu, Kun Zhao, Hongxia Yang, Wei Lin, Chang Zhou, Baole Ai, Yong Li, and Jingren Zhou. 2019. AliGraph: A Comprehensive Graph Neural Network Platform. In *PVLDB*.
38. Peter Ahrens and Erik G. Boman. 2020. On Optimal Partitioning For Sparse Matrices In Variable Block Row Format. *CoRR* abs/2005.12414 (2020). https://arxiv.org/abs/2005.12414
39. Réka Albert, Hawoong Jeong, and Albert-László Barabási. 1999. Diameter of the world-wide web. *nature* 401, 6749 (1999), 130–131.
40. Iz Beltagy, Matthew E. Peters, and Arman Cohan. 2020. Longformer: The Long-Document Transformer. *CoRR* abs/2004.05150 (2020). https://arxiv.org/abs/2004.05150
41. Beidi Chen, Tri Dao, Kaizhao Liang, Jiaming Yang, Zhao Song, Atri Rudra, and Christopher Ré. 2022. Pixelated Butterfly: Simple and Efficient Sparse training for Neural Network Models. In *The Tenth International Conference on Learning Representations, ICLR 2022, Virtual Event, April 25-29, 2022*. OpenReview.net. https://openreview.net/forum?id=Nfl-iXa-y7R
42. Zhaodong Chen, Zheng Qu, Liu Liu, Yufei Ding, and Yuan Xie. 2021. Efficient tensor core-based gpu kernels for structured sparsity under reduced precision. In *Proceedings of the International Conference for High Performance Computing, Networking, Storage and Analysis*. 1–14.
43. Rewon Child, Scott Gray, Alec Radford, and Ilya Sutskever. 2019a. Generating long sequences with sparse transformers. *arXiv preprint* arXiv:1904.10509 (2019).
44. Rewon Child, Scott Gray, Alec Radford, and Ilya Sutskever. 2019b. Generating Long Sequences with Sparse Transformers. *CoRR* abs/1904.10509 (2019). http://arxiv.org/abs/1904.10509

45. JeeWhan Choi, Amik Singh, and Richard W. Vuduc. 2010. Model-driven autotuning of sparse matrix-vector multiply on GPUs. In *Proceedings of the 15th ACM SIGPLAN Symposium on Principles and Practice of Parallel Programming, PPOPP 2010, Bangalore, India, January 9-14, 2010*, R. Govindarajan, David A. Padua, and Mary W. Hall (Eds.). ACM, 115–126. https://doi.org/10.1145/1693453.1693471
46. Stephen Chou, Fredrik Kjolstad, and Saman Amarasinghe. 2018. Format abstraction for sparse tensor algebra compilers. *Proceedings of the ACM on Programming Languages* 2, OOPSLA (2018), 1–30.
47. NVIDIA Corporation. 2022. *cuSPARSE :: CUDA Toolkit Documentation v11.7.1*. Retrieved July 15, 2023 from https://docs.nvidia.com/cuda/cusparse/index.html
48. Zhen Du, Jiajia Li, Yinshan Wang, Xueqi Li, Guangming Tan, and Ninghui Sun. 2022. AlphaSparse: Generating High Performance SpMV Codes Directly from Sparse Matrices. In *SC22: International Conference for High Performance Computing, Networking, Storage and Analysis, Dallas, TX, USA, November 13-18, 2022*. IEEE, 1–15. https://doi.org/10.1109/SC41404.2022.00071
49. Trevor Gale, Matei Zaharia, Cliff Young, and Erich Elsen. 2020a. Sparse GPU kernels for deep learning. In *Proceedings of the International Conference for High Performance Computing, Networking, Storage and Analysis, SC 2020, Virtual Event / Atlanta, Georgia, USA, November 9-19, 2020*, Christine Cuicchi, Irene Qualters, and William T. Kramer (Eds.). IEEE/ACM, 17. https://doi.org/10.1109/SC41405.2020.00021
50. Trevor Gale, Matei Zaharia, Cliff Young, and Erich Elsen. 2020b. Sparse GPU kernels for deep learning. In *SC20: International Conference for High Performance Computing, Networking, Storage and Analysis*. IEEE, 1–14.
51. Will Hamilton, Zhitao Ying, and Jure Leskovec. 2017. Inductive representation learning on large graphs. *Advances in neural information processing systems* 30 (2017).
52. Song Han, Jeff Pool, John Tran, and William J. Dally. 2015. Learning both Weights and Connections for Efficient Neural Networks. *CoRR* abs/1506.02626 (2015). arXiv:1506.02626 http://arxiv.org/abs/1506.02626
53. Refael Hassin and Asaf Levin. 2005. A Better-Than-Greedy Approximation Algorithm for the Minimum Set Cover Problem. *SIAM J. Comput.* 35, 1 (2005), 189–200. https://doi.org/10.1137/S0097539704444750
54. Ahmed E. Helal, Jan Laukemann, Fabio Checconi, Jesmin Jahan Tithi, Teresa M. Ranadive, Fabrizio Petrini, and Jeewhan Choi. 2021. ALTO: adaptive linearized storage of sparse tensors. In *ICS '21: 2021 International Conference on Supercomputing, Virtual Event, USA, June 14-17, 2021*, Huiyang Zhou, Jose Moreira, Frank Mueller, and Yoav Etsion (Eds.). ACM, 404–416. https://doi.org/10.1145/3447818.3461703
55. Changwan Hong, Aravind Sukumaran-Rajam, Israt Nisa, Kunal Singh, and P. Sadayappan. 2019. Adaptive sparse tiling for sparse matrix multiplication. In *Proceedings of the 24th ACM SIGPLAN Symposium on Principles and Practice of Parallel Programming, PPoPP 2019, Washington, DC, USA, February 16-20, 2019*, Jeffrey K. Hollingsworth and Idit Keidar (Eds.). ACM, 300–314. https://doi.org/10.1145/3293883.3295712
56. Yuwei Hu, Zihao Ye, Minjie Wang, Jiali Yu, Da Zheng, Mu Li, Zheng Zhang, Zhiru Zhang, and Yida Wang. 2020b. FeatGraph: a flexible and efficient backend for graph neural network systems. In *Proceedings of the International Conference for High Performance Computing, Networking, Storage and Analysis, SC 2020, Virtual Event / Atlanta, Georgia, USA, November 9-19, 2020*, Christine Cuicchi, Irene Qualters, and William T. Kramer (Eds.). IEEE/ACM, 71. https://doi.org/10.1109/SC41405.2020.00075
57. Paolo Sylos Labini, Massimo Bernaschi, Francesco Silvestri, and Flavio Vella. 2022. Blocking Techniques for Sparse Matrix Multiplication on Tensor Accelerators. *CoRR* abs/2202.05868 (2022). arXiv:2202.05868 https://arxiv.org/abs/2202.05868
58. François Lagunas, Ella Charlaix, Victor Sanh, and Alexander M. Rush. 2021. Block Pruning For Faster Transformers. In *Proceedings of the 2021 Conference on Empirical Methods in Natural Language Processing, EMNLP 2021, Virtual Event / Punta Cana, Dominican Republic, 7-11 November, 2021*, Marie-Francine Moens, Xuanjing Huang, Lucia Specia, and

Scott Wen-tau Yih (Eds.). Association for Computational Linguistics, 10619–10629. https://doi.org/10.18653/v1/2021.emnlp-main.829
59. Ao Li, Bojian Zheng, Gennady Pekhimenko, and Fan Long. 2022b. Automatic Horizontal Fusion for GPU Kernels. In *IEEE/ACM International Symposium on Code Generation and Optimization, CGO 2022, Seoul, Korea, Republic of, April 2-6, 2022*, Jae W. Lee, Sebastian Hack, and Tatiana Shpeisman (Eds.). IEEE, 14–27. https://doi.org/10.1109/CGO53902.2022.9741270
60. Shiyang Li, Xiaoyong Jin, Yao Xuan, Xiyou Zhou, Wenhu Chen, Yu-Xiang Wang, and Xifeng Yan. 2019. Enhancing the locality and breaking the memory bottleneck of transformer on time series forecasting. *Advances in neural information processing systems* 32 (2019).
61. Shigang Li, Kazuki Osawa, and Torsten Hoefler. 2022a. Efficient Quantized Sparse Matrix Operations on Tensor Cores. In *SC22: International Conference for High Performance Computing, Networking, Storage and Analysis, Dallas, TX, USA, November 13-18, 2022*, Felix Wolf, Sameer Shende, Candace Culhane, Sadaf R. Alam, and Heike Jagode (Eds.). IEEE, 37:1–37:15. https://doi.org/10.1109/SC41404.2022.00042
62. Weifeng Liu and Brian Vinter. 2015. CSR5: An efficient storage format for cross-platform sparse matrix-vector multiplication. In *Proceedings of the 29th ACM on International Conference on Supercomputing*. 339–350.
63. JS McCarley, Rishav Chakravarti, and Avirup Sil. 2019. Structured pruning of a bert-based question answering model. *arXiv preprint* arXiv:1910.06360 (2019).
64. Atefeh Mehrabi, Donghyuk Lee, Niladrish Chatterjee, Daniel J. Sorin, Benjamin C. Lee, and Mike O'Connor. 2021. Learning Sparse Matrix Row Permutations for Efficient SpMM on GPU Architectures. In *IEEE International Symposium on Performance Analysis of Systems and Software, ISPASS 2021, Stony Brook, NY, USA, March 28-30, 2021*. IEEE, 48–58. https://doi.org/10.1109/ISPASS51385.2021.00016
65. Yuyao Niu, Zhengyang Lu, Meichen Dong, Zhou Jin, Weifeng Liu, and Guangming Tan. 2021. TileSpMV: A Tiled Algorithm for Sparse Matrix-Vector Multiplication on GPUs. In *35th IEEE International Parallel and Distributed Processing Symposium, IPDPS 2021, Portland, OR, USA, May 17-21, 2021*. IEEE, 68–78. https://doi.org/10.1109/IPDPS49936.2021.00016
66. Yuyao Niu, Zhengyang Lu, Haonan Ji, Shuhui Song, Zhou Jin, and Weifeng Liu. 2022. TileSpGEMM: a tiled algorithm for parallel sparse general matrix-matrix multiplication on GPUs. In *PPoPP '22: 27th ACM SIGPLAN Symposium on Principles and Practice of Parallel Programming, Seoul, Republic of Korea, April 2 - 6, 2022*, Jaejin Lee, Kunal Agrawal, and Michael F. Spear (Eds.). ACM, 90–106. https://doi.org/10.1145/3503221.3508431
67. Weihua Hu, Matthias Fey, Marinka Zitnik, Yuxiao Dong, Hongyu Ren, Bowen Liu, Michele Catasta, and Jure Leskovec. 2020. Open graph benchmark: Datasets for machine learning on graphs. *NeurIPS* 33 (2020), 22118–22133. *NeurIPS* 33 (2020), 22118–22133.
68. Victor Sanh, Thomas Wolf, and Alexander M. Rush. 2020. Movement Pruning: Adaptive Sparsity by Fine-Tuning. In *Advances in Neural Information Processing Systems 33: Annual Conference on Neural Information Processing Systems 2020, NeurIPS 2020, December 6-12, 2020, virtual*, Hugo Larochelle, Marc'Aurelio Ranzato, Raia Hadsell, Maria-Florina Balcan, and Hsuan-Tien Lin (Eds.). https://proceedings.neurips.cc/paper/2020/hash/eae15aabaa768ae4a5993a8a4f4fa6e4-Abstract.html
69. Prithviraj Sen, Galileo Namata, Mustafa Bilgic, Lise Getoor, Brian Galligher, and Tina Eliassi-Rad. 2008. Collective classification in network data. *AI magazine* 29, 3 (2008), 93–93.
70. Philippe Tillet, Hsiang-Tsung Kung, and David D. Cox. 2019. Triton: an intermediate language and compiler for tiled neural network computations. In *Proceedings of the 3rd ACM SIGPLAN International Workshop on Machine Learning and Programming Languages, MAPL@PLDI 2019, Phoenix, AZ, USA, June 22, 2019*, Tim Mattson, Abdullah Muzahid, and Armando Solar-Lezama (Eds.). ACM, 10–19. https://doi.org/10.1145/3315508.3329973
71. Minjie Wang, Da Zheng, Zihao Ye, Quan Gan, Mufei Li, Xiang Song, Jinjing Zhou, Chao Ma, Lingfan Yu, Yu Gai, et al. 2019. Deep graph library: A graph-centric, highly-performant package for graph neural networks. *arXiv preprint* arXiv:1909.01315 (2019).

72. Yuke Wang, Boyuan Feng, and Yufei Ding. 2021a. TC-GNN: Accelerating Sparse Graph Neural Network Computation Via Dense Tensor Core on GPUs. *CoRR* abs/2112.02052 (2021). arXiv:2112.02052 https://arxiv.org/abs/2112.02052
73. Yang Wang, Chen Zhang, Zhiqiang Xie, Cong Guo, Yunxin Liu, and Jingwen Leng. 2021b. Dual-side Sparse Tensor Core. In *48th ACM/IEEE Annual International Symposium on Computer Architecture, ISCA 2021, Valencia, Spain, June 14-18, 2021*. IEEE, 1083–1095. https://doi.org/10.1109/ISCA52012.2021.00088
74. Samuel Williams, Andrew Waterman, and David A. Patterson. 2009. Roofline: an insightful visual performance model for multicore architectures. *Commun. ACM* 52, 4 (2009), 65–76. https://doi.org/10.1145/1498765.1498785
75. Jaeyeon Won, Charith Mendis, Joel S Emer, and Saman Amarasinghe. 2023. WACO: Learning Workload-Aware Co-optimization of the Format and Schedule of a Sparse Tensor Program. In *Proceedings of the 28th ACM International Conference on Architectural Support for Programming Languages and Operating Systems, Volume 2*. 920–934.
76. Zihao Ye, Ruihang Lai, Junru Shao, Tianqi Chen, and Luis Ceze. 2023. SparseTIR: Composable Abstractions for Sparse Compilation in Deep Learning. In *Proceedings of the 28th ACM International Conference on Architectural Support for Programming Languages and Operating Systems, Volume 3, ASPLOS 2023, Vancouver, BC, Canada, March 25–29, 2023*, Tor M. Aamodt, Natalie D. Enright Jerger, and Michael M. Swift (Eds.). ACM, 660–678. https://doi.org/10.1145/3582016.3582047
77. Zhongming Yu, Guohao Dai, Guyue Huang, Yu Wang, and Huazhong Yang. 2021. Exploiting Online Locality and Reduction Parallelism for Sampled Dense Matrix Multiplication on GPUs. In *39th IEEE International Conference on Computer Design, ICCD 2021, Storrs, CT, USA, October 24-27, 2021*. IEEE, 567–574. https://doi.org/10.1109/ICCD53106.2021.00092
78. Ningxin Zheng, Bin Lin, Quanlu Zhang, Lingxiao Ma, Yuqing Yang, Fan Yang, Yang Wang, Mao Yang, and Lidong Zhou. 2022. SparTA:Deep-Learning Model Sparsity via Tensor-with-Sparsity-Attribute. In *16th USENIX Symposium on Operating Systems Design and Implementation (OSDI 22)*. 213–232.
79. Yangjie Zhou, Jingwen Leng, Yaoxu Song, Shuwen Lu, Mian Wang, Chao Li, Minyi Guo, Wenting Shen, Yong Li, Wei Lin, et al. 2023. uGrapher: High-Performance Graph Operator Computation via Unified Abstraction for Graph Neural Networks. In *Proceedings of the 28th ACM International Conference on Architectural Support for Programming Languages and Operating Systems, Volume 2*. 878–891.

Chapter 5
Distributed Graph Neural Network Training

Jingzhi Fang and Jingshu Peng

Abstract In this chapter, we present our distributed GNN training solution. Specifically, to mitigate the communication overhead induced by expensive data movement across workers, we propose Sancus, a staleness-aware, communication-avoiding, decentralized GNN system. By introducing a set of novel bounded embedding staleness metrics and adaptively skipping broadcasts, Sancus abstracts decentralized GNN computation as sequential matrix multiplication and leverages cached historical embeddings. Theoretically, we establish bounded approximation errors for both embeddings and gradients, together with convergence guarantees. Empirically, we evaluate Sancus with common GNN models under diverse system configurations on large-scale benchmark datasets. Compared to SOTA works, Sancus avoids up to 74% communication and achieves at least 1.86× higher throughput on average, without accuracy loss.

5.1 Background

The success of Graph Neural Networks (GNNs) [1] has driven recent advances in state-of-the-art modeling for real-world graphs. At a high level, GNNs are structure-aware models whose architectures are naturally aligned with the underlying topology of the input graph. By repeatedly aggregating information from target nodes' neighborhoods, GNNs jointly capture structural dependencies and node features. Despite their strong performance, scaling GNNs to large graphs remains challenging due to the difficulty of processing all data within a finite time budget and the limited scalability of existing algorithms. To alleviate the memory requirement under ever-growing data and model sizes, distributed GNN processing becomes an inevitable choice. Prior work has explored sampling-based distributed GNN training, but often at the cost of information loss [2–4], additional sampling overhead [3], and even the

J. Fang (✉) · J. Peng
The Hong Kong University of Science and Technology, Hong Kong SAR, China
e-mail: jfangak@cse.ust.hk

J. Peng
e-mail: jpengab@cse.ust.hk

Y. Shen and L. Chen (eds.), *Graph Neural Network Training*, Machine Learning: Foundations, Methodologies, and Applications,
https://doi.org/10.1007/978-981-95-5795-0_5

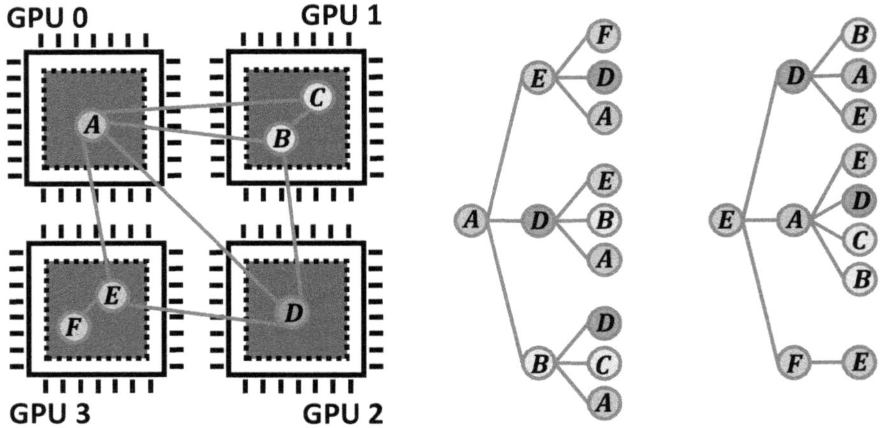

Fig. 5.1 Distributed full-GNN example: nodes in same color are on same GPU. On the left, a 6-node graph is stored on 4 GPUs. On the right are the computational graphs of a 2-layer GNN for node A and E. During GNN neighborhood aggregation, intensive cross-device visits (10 times to update node A and 9 for E) to fetch neighbor data cause expensive communication overhead

lack of convergence guarantees [5]. Therefore, in this work, we focus on distributed training for full-GNNs.

Compared to traditional graph processing or machine learning, distributed full-GNNs introduce new system-level challenges. Beyond their substantial memory footprint, distributed GNNs are both memory-intensive and compute-intensive [6, 7], since irregular neighborhood accesses are tightly coupled with an iterative learning procedure. As a result, heavy communication becomes a key obstacle to efficiency: in addition to gradients and parameters, workers must frequently exchange intermediate embeddings. As illustrated in Fig. 5.1, training continuously queries target nodes, their neighbors, and multi-hop neighborhoods, requiring embeddings and gradients to be transferred across workers throughout the process. Therefore, due to the inherent data movement in GNN aggregation, cross-device data communication becomes a major archenemy of efficient GNN processing. In practice, the resulting communication overhead can account for 80% or more of the total training time [2, 4, 8].

In distributed training, the communication architecture among workers is pivotal, especially for GNNs where communication overhead is substantial. As shown in Fig. 5.2, distributed systems typically adopt either a centralized or a decentralized design. Most existing distributed GNN systems [3, 8–12] follow the centralized parameter server (PS) paradigm in Fig. 5.2a. To achieve efficiency and scalability, they often rely on heavy preprocessing and a complex execution workflow. However, due to the intrinsic communication demands of GNNs, frequent exchanges between all workers and the central PS, together with synchronization delays caused by stragglers, can result in high communication overhead [2]. In contrast, decentralized archi-

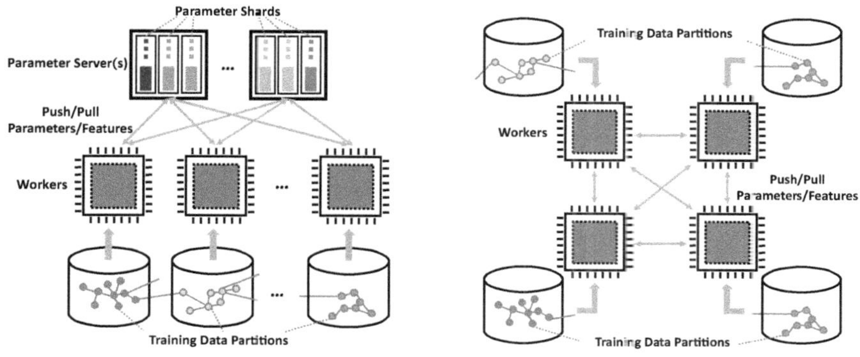

(a) Centralized. (b) Decentralized.

Fig. 5.2 Centralized versus decentralized GNN processing example. In the centralized scheme, workers periodically send updates to a parameter server. In the decentralized scheme, workers exchange them directly

tectures are often more robust and easier to deploy: they avoid the engineering burden of implementing and tuning a PS, mitigate centralized bandwidth bottlenecks, and eliminate a single point of failure [13]. Moreover, for large neural networks, decentralized schemes are theoretically shown to be more advantageous [14]. Motivated by these benefits, Tripathy et al. [4] propose CAGNET, which (to our knowledge) is the only SOTA decentralized parallel algorithm [15] tailored to GNN training. In CAGNET, each worker maintains a full replica of the model parameters to reduce centralized communication. Nevertheless, it still incurs redundant and unnecessary broadcasts of embeddings and gradients. In addition, workers must synchronize and wait for stragglers, which further amplifies the communication overhead.

In this chapter, to address this gap in efficient GNN processing, we propose Sancus, a staleness-aware, communication-avoiding, decentralized GNN training system that adaptively skips broadcasts and caches historical embeddings under bounded staleness. To bypass irregular inter-GPU data communication, we firstly revisit parallel algorithms for distributed GNN training [4] and reduce communication overhead in a fundamentally different manner. As Fig. 5.3 shows, by treating decentralized GNN computation purely as a sequence of matrix multiplication operations, each GPU loads partitioned submatrices without relying on graph-specific semantics. Specifically, the (often prohibitively large) adjacency and embedding matrices are partitioned into $\mathbf{A}_i$ and $\mathbf{H} * i$ $(i \in [1, 4])$ and placed on GPU$*i - 1$, together with a full replica of the weight matrix W. Then $\mathbf{H}_1$ through $\mathbf{H}_4$ are broadcast sequentially in a one-to-all manner to all GPUs. After 4 such broadcasts, the full embedding matrix $\mathbf{H}$ is updated for one layer. By moving intact matrix blocks, Sancus exploits data parallelism to avoid communication that would otherwise arise from intensive neighbor fetching in Fig. 5.1. Secondly, to further avoid communication in the decentralized setting, we cache and skip-broadcast historical embeddings. We define historical embeddings as embedding submatrices computed in earlier epochs on each worker,

Table 5.1 Summary of the key symbols and notations

Notation	Description
$\mathbf{A}$	Adjacency matrix of the graph ($N \times N$)
$\hat{\mathbf{A}}$	Adjacency matrix after symmetric normalization ($N \times N$)
$\mathbf{W}^{(\ell)}$	Weight matrix at the ℓth layer ($F \times F$)
$\mathbf{H}^{(\ell)}$	Embedding matrix at the ℓth layer ($N \times F$)
$\mathbf{Z}^{(\ell)}$	Input matrix to activation function at the ℓth layer ($N \times F$)
$\mathbf{T}^{(\ell)}$	Intermediate result matrix of the multiplication $\hat{\mathbf{A}}\mathbf{H}^{(\ell)}$
$\nabla_{\mathbf{Z}^{(\ell)}}\mathcal{L} = \delta^{(\ell)}$	Gradient matrix of the loss $\mathcal{L}$ with respect to $\mathbf{Z}^{(\ell)}$ at the ℓth layer
$\nabla_{\mathbf{W}^{(\ell)}}\mathcal{L}$	Gradient matrix of the loss $\mathcal{L}$ with respect to $\mathbf{W}^{(\ell)}$ at the ℓth layer
$\mathcal{L}$	Loss of the GNN
η	Learning rate
ϵ	Staleness bound
$P(i)$	ith process in distributed GNN

i.e., the submatrix $\mathbf{H} * i$ produced locally on GPU$*i - 1$ in Fig. 5.3. Based on this observation, we leverage caching and design a novel skip-broadcast operator in Sancus to enable the reuse of historical embeddings. Thirdly, to control the system staleness introduced by using mixed-version embeddings across GPUs, we generalize the widely used bounded-gradient staleness in centralized settings [16] to historical embeddings, and introduce a set of novel bounded embedding staleness metrics for decentralized GNNs. Concretely, Sancus adaptively skip-broadcasts embeddings as long as their staleness remains within prescribed bounds, thereby reusing cached historical embeddings to avoid communication directly; once embeddings become too stale, the latest results are broadcast and the cache is refreshed to keep the system staleness bounded. Using Fig. 5.3 as a toy example, if $\mathbf{H}_3$ is not too stale, Sancus can skip one broadcast; in this case, Sancus needs only 7 broadcasts to update all embeddings for a 2-layer GNN. We emphasize again that Sancus eliminates per-node embedding fetching during aggregation. In contrast, for the conventional distributed GNN in Fig. 5.1, obtaining the embeddings for nodes A and E alone requires 10 and 9 request-and-send operations, respectively; updating all nodes requires 59 such operations. However, as shown in Fig. 5.3, Sancus completes the same update with as few as 7 broadcasts in total. Finally, Sancus imposes little preprocessing burden and can be readily applied to distributed GNN training based on arbitrary matrix blocking and direct matrix operations.

Following the definitions of GNNs and GNN training in Sect. 2.2.1, we formally define the problem in this chapter. The key notations are summarized in Table 5.1. We focus on node-level graph representation learning with distributed GNNs.

Definition 5.1 (*Node-level Graph Representation Learning*) Given an L-layer GNN model, an undirected graph $\mathcal{G} = (\mathcal{V}, \mathcal{E})$ with adjacency matrix $\mathbf{A}$, an initial feature matrix $\mathbf{X}$, and labels $\mathbf{Y}$, the representation matrix $\mathbf{H}^{(L)}$ consists of learned low-dimensional vector representations for all nodes. Formally, given a loss function $\mathcal{L}$

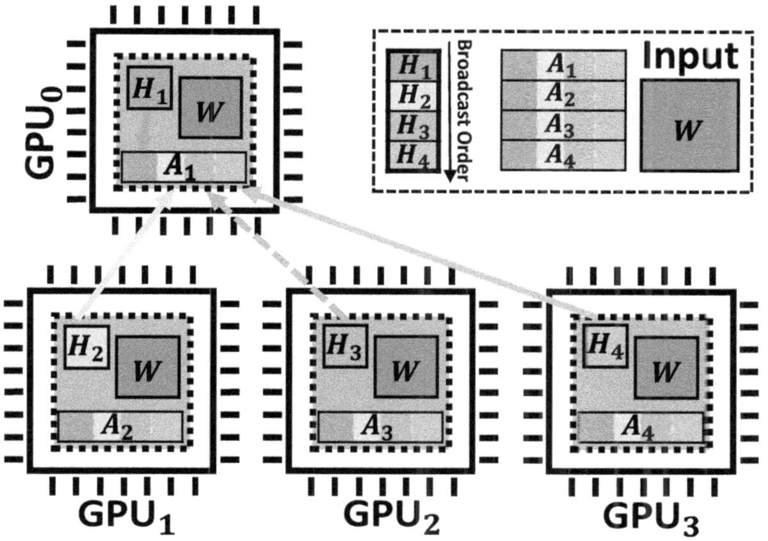

Fig. 5.3 Toy 2-layer GNN example under Sancus. Each GPU$_{i-1}$ stores its local shards of $\mathbf{H}_i$ and $\mathbf{A}_i$ while holding a full copy of W. At each layer, the embedding blocks $\mathbf{H}i$ are disseminated to all GPUs sequentially via one-to-all broadcast (arrows omitted where unambiguous). After four broadcasts $\rightarrow$, $\rightarrow$, $\rightarrow$, $\rightarrow$ and local computation, the broadcasts proceed in a pipelined parallel fashion, updating the full embedding view needed by every GPU$_{i-1}$. In the next layer, Sancus can deem $\mathbf{H}_3$ sufficiently fresh and skip one broadcast, indicated by $\dashrightarrow$. Overall, the 2-layer update requires only $4 + 3 = 7$ broadcasts

and targets $\mathbf{Y}$, we solve the following objective:

$$
\begin{aligned}
&\min_{\mathbf{W}} \ \mathcal{L}\left(\mathbf{H}^{(L)}, \mathbf{Y}\right) \\
&\text{s.t.} \ \ \mathbf{H}^{(\ell)} = \sigma\left(\mathbf{H}^{(\ell-1)}, \hat{\mathbf{A}}; \mathbf{W}^{(\ell-1)}\right), \ \forall \ell \in [1, L].
\end{aligned}
$$

5.2 The Sancus Framework

First, we present a step-by-step overview of Sancus in Sect. 5.2.1. Algorithm 1 describes the complete staleness-aware, communication-avoiding, decentralized full-GNN training algorithm. To further detail how communication is avoided, we introduce historical embeddings and the corresponding skip-broadcast mechanism in Sects. 5.2.3 and 5.2.4. To control the system staleness induced by historical embeddings, we propose a set of novel bounded embedding staleness metrics in Sect. 5.2.5.

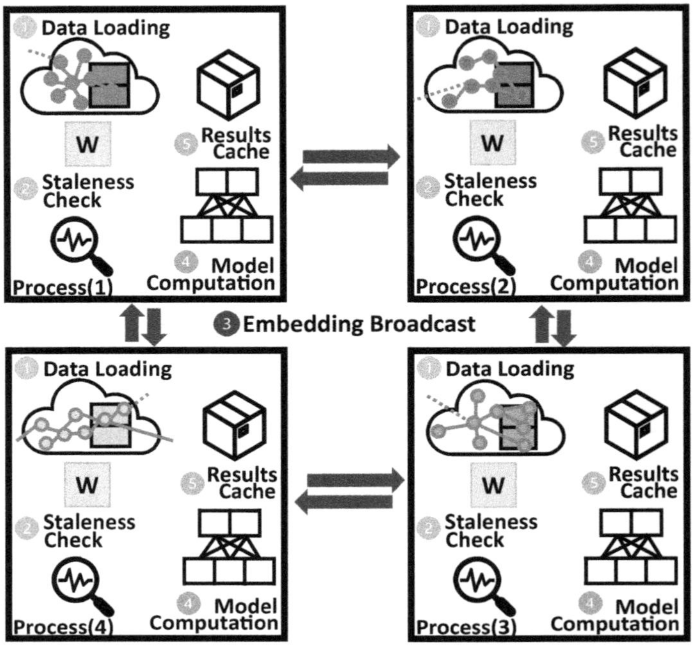

Fig. 5.4 Overall architecture. The full graph and node features are partitioned and assigned to individual processes across workers. Sancus proceeds in five stages. (1) Data loading: each GPU loads its graph partition, the corresponding embedding-matrix blocks, and a full replica of the model; (2) staleness check: for the current root GPU, the system evaluates whether its embedding block violates the staleness bound; (3) embedding broadcast: if the block is deemed too stale, the root broadcasts the latest embeddings to all GPUs; otherwise, other GPUs reuse their cached historical copy associated with the current root; (4) model computation: each GPU runs the GNN computation using either fresh embeddings or cached stale embeddings; (5) results cache: caches are refreshed as needed with newly received embeddings

5.2.1 *Overview*

In this work, we propose Sancus, an adaptive, staleness-aware, communication-avoiding, decentralized GNN system. Fundamentally, Sancus is simple yet effective: it caches and reuses stale historical embeddings and accordingly skips broadcasts during decentralized GNN training, building on a general communication-avoiding matrix blocking algorithm for parallel computing.

We provide an overview of Sancus in Fig. 5.4. Overall, Sancus consists of five steps: (1) data loading, (2) staleness bound checking, (3) embedding broadcasting, (4) GNN model computing, and (5) results caching. We briefly explain these steps as follows.

(1) first, the sparse adjacency matrix of the full graph and the dense embedding matrix are partitioned into matrix blocks and loaded onto individual workers. Each worker maintains its own replica of the full model;
(2) on each GPU, before broadcasting the results from the previous computation, we check whether the staleness of the historical embeddings remains within the proposed bounds. If so, the embedding broadcast is skipped and cached historical embeddings are reused for model computation in this iteration;
(3) otherwise, if the staleness exceeds the limit, the latest results are broadcast to all workers and the cache is updated accordingly;
(4) consequently, the GNN model computes using either the latest embeddings or the cached historical embeddings;
(5) finally, the updated embeddings are passed to the next iteration for staleness checking prior to any broadcast.

5.2.2 Staleness-Aware Communication-Avoiding Decentralized Training

First, we present the complete staleness-aware, communication-avoiding, decentralized full-graph GNN training procedure in Algorithm 1 and elaborate on its details. It hinges on three key components:

(1) a worker-state flag $F(i)$ that indicates the state of worker i. Each worker is marked as either ACTIVE or STALE, enabling the Skip-Broadcast operation in Sect. 5.2.4;
(2) a cache that stores historical embeddings received from other workers, which can be repeatedly reused in future iterations to avoid communication;
(3) tolerance of bounded embedding staleness to control system staleness, where different workers may consume embeddings computed in different iterations.

During preprocessing, Sancus is compatible with a broad range of partitioners that decompose the graph-related matrices into blocks. This includes widely used choices such as METIS [17] (employed by systems like DistDGL [11] and AliGraph [12]), as well as lightweight alternatives such as random partitioning. As depicted in Fig. 5.5, Sancus deliberately casts full-graph GNN computation into a sequence of matrix multiplications, thereby sidestepping the heavy neighbor-fetching pattern typically incurred by aggregation. Concretely, the sparse adjacency matrix $\hat{\mathbf{A}}$ and the dense embedding matrix $\mathbf{H}$ are block-distributed across worker processes $P(i)$, $i \in [1, p]$. Let N be the number of nodes and F the embedding dimension. Then $\hat{\mathbf{A}}$ has shape $(N \times N)$ and is partitioned along both rows and columns into $p \times p$ blocks, whereas $\mathbf{H}$ has shape $(N \times F)$ and is partitioned only by rows into p blocks, as shown in Fig. 5.5. In contrast, the $(F \times F)$ weight matrix $\mathbf{W}$ remains dense and is replicated in every process $P(i)$. Furthermore, for the multiplication $\hat{\mathbf{A}}\mathbf{H}^{(\ell)}$, we denote the per-process partial result on $P(i)$ by $\mathbf{T} * i^{(\ell)}$, defined as $\mathbf{T} * i^{(\ell)} = \sum * j = 1^p \hat{\mathbf{A}} * ij \mathbf{H}_j^{(\ell)}$.

Algorithm 1 The decentralized stale parallel GNN training algorithm based on arbitrary general block row decomposition preprocessing strategy with a foward pass procedure to compute $\mathbf{Z}$ in Eq. (2.3) and $\mathbf{H}$ in Eq. (2.4), a backpropagation procedure to compute the gradients δ in Eq. (2.5) and $\nabla_{\mathbf{W}}\mathcal{L}$ in Eq. (2.6), and the final weight update in Eq. (2.7). The matrices $\hat{\mathbf{A}}$ and $\mathbf{H}$ are distributed on a $p \times p$ process grid, where each process $P(i)$ receives N/p consecutive block rows

Input: $\mathcal{G} = (\mathcal{V}, \mathcal{E})$; Sparse adjacency matrix $\hat{\mathbf{A}}$; Dense feature matrix $\mathbf{H}^{(0)}$; Dense weight matrix $\mathbf{W}$; Output: Node embedding matrix $\mathbf{H}^{(L)}$;

1: Preprocessing: Block row partition;
2: **for all** process $P(i)$ in parallel **do**
3: **procedure** FORWARD PASS
4: **for** $\ell = 1, \cdots, L$ **do**
5: **for** $j = 1$ to p **do**
6: **if** $F(j) == ACTIVE$ **then** ▷ Worker state flag
7: BROADCAST$\mathbf{H}_j^{(\ell-1)}$
8: CACHE$\mathbf{H}_j^{(\ell-1)}$ ▷ Cache latest $\mathbf{H}_j^{(\ell-1)}$ from $P(j)$
9: ▷ Compute intermediates with latest $\mathbf{H}_j^{(\ell-1)}$
10: $\mathbf{T}_i^{(\ell-1)} \leftarrow \mathbf{T}_i^{(\ell-1)} + \hat{\mathbf{A}}_{ij}\mathbf{H}_j^{(\ell-1)}$
11: **else** \\ Skip-Broadcast
12: ▷ Compute intermediates with historical $\tilde{\mathbf{H}}_j^{(\ell-1)}$
13: $\mathbf{T}_i^{(\ell-1)} \leftarrow \mathbf{T}_i^{(\ell-1)} + \hat{\mathbf{A}}_{ij}\tilde{\mathbf{H}}_j^{(\ell-1)}$
14: **end if**
15: $\mathbf{Z}_i^{(\ell)} \leftarrow \mathbf{T}_i^{(\ell-1)}\mathbf{W}^{(\ell-1)}$ ▷ Input to activation
16: $\mathbf{H}_i^{(\ell)} \leftarrow \sigma\left(\mathbf{Z}_i^{(\ell)}\right)$ ▷ Update the embeddings
17: $F(i) \leftarrow$ STALE$\mathbf{H}_i^{(\ell)}$ ▷ Update state of worker i
18: **end for**
19: **end for**
20: **procedure** BACKWARD PASS
21: **for** $\ell = L-1, \cdots, 0$ **do**
22: $\delta_i^{(\ell)} \leftarrow$ GRADIENT_CLIP$\delta_i^{(\ell)}$
23: BROADCAST$\delta_i^{(\ell)}$ and Update weights $\mathbf{W}^{\ell}$ by gradients
24: **end for**
25: **end procedure**
26: **end procedure**
27: **end for**

Across all distributed processes $P(i)$, Sancus runs forward computation and backpropagation concurrently, relying on standard collectives (e.g., ring-based pipelined Broadcast and AllReduce) for coordination. At the start of each communication phase, process j evaluates the staleness of its intermediate embedding block $\mathbf{H} * j^{(\ell-1)}$ (Line 6) before deciding whether it should be disseminated. When the worker is marked ACTIVE, $\mathbf{H} * j^{(\ell-1)}$ is pushed out from the root rank and propagated to all ranks via a sequential One-to-All Broadcast (Line 7); the received block is then stored in the cache (Line 8). If the worker is instead STALE, Sancus invokes Skip-Broadcast: process j is removed from the sender side of the current communication

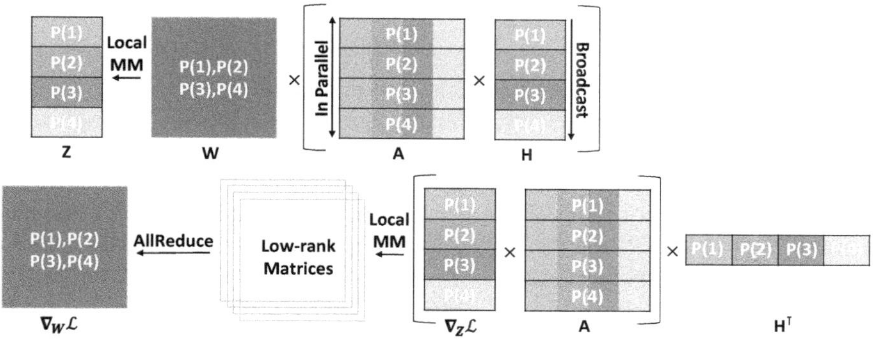

Fig. 5.5 Communication-avoiding data-parallel GNN training, from a sequential matrix multiplication processing perspective

schedule, while still remaining a receiver in the broadcast graph so that it can obtain others' updates. In this round, j does not transmit $\mathbf{H} * j^{(\ell-1)}$; consequently, every other worker falls back to its cached (possibly outdated) copy $\tilde{\mathbf{H}} * j^{(\ell-1)}$. In other words, local computation proceeds using either the latest block from the previous epoch (Line 10) or an automatically reused cached block from earlier epochs (Line 13). With the required blocks available, each process $P(i)$ forms its intermediate product $\mathbf{T} * i^{(\ell-1)}$ and computes the next-layer embeddings $\mathbf{H} * i^{(\ell)}$. After obtaining $\mathbf{H} * i^{(\ell)}$, process i updates its worker-state flag $F(i)$ based on whether the new block satisfies the staleness bound (Line 17): it stays STALE to continue skipping broadcasts of $\mathbf{H} * i^{(\ell)}$, or switches to ACTIVE so that $\mathbf{H}_i^{(\ell)}$ will be broadcast in subsequent iterations. Depending on the configuration, this check follows Definitions 5.2, 5.3, or 5.4 in Line 17.

During backpropagation, gradient blocks δ_j follow an analogous dissemination pattern. To curb the amount of gradient communication, Sancus applies local gradient clipping in Line 22, which also acts as a regularizer when historical embeddings are reused. In standard practice, clipping is triggered when the L2 norm $||\cdot||_2$ exceeds a threshold th. Under a decentralized setup with P workers, we use a per-worker threshold th_i. Suppose gradients on different workers are i.i.d. with variance σ^2; then aggregating gradients across workers yields variance $P\sigma^2$. As a consequence, $\mathbb{E}||\delta_i||_2 \approx \sigma$ while $\mathbb{E}||\delta||_2 \approx P^{\frac{1}{2}}\sigma$. This motivates scaling the local clipping threshold by $P^{\frac{1}{2}}$, i.e., $\text{th}_i = P^{\frac{1}{2}}\text{th}$. Finally, model updates are performed via an AllReduce operation, which aggregates gradients from all workers and distributes the combined result to every rank.

By modeling GNN computation purely as a sequence of matrix multiplications (cf. Fig. 5.5), Sancus communicates only coarse-grained matrix blocks among decentralized workers. This design eliminates the irregular, fine-grained request–send messaging pattern used to fetch neighbors in vertex-centric distributed GNN frameworks. To further reduce communication, we introduce the notion of historical embeddings in Sect. 5.2.3, enabling Sancus to cache and reuse previously computed embedding blocks.

5.2.3 Historical Embeddings

Assume a decentralized setup with P workers. We partition the embedding matrix $\mathbf{H}$ by rows into blocks $\mathbf{H}_i$, where $i \in [1, p]$, and assign each block to the corresponding process $P(i)$. To form the layer-ℓ embedding $\mathbf{H}^{(\ell)}$ of a general GNN in Eq. (2.2), workers must conceptually assemble contributions from all embedding blocks across GPUs. Motivated by prior uses of historical embeddings $\tilde{\mathbf{h}}^{(\ell)}$ [5, 18], we extend this notion to distributed GNN training by treating the intermediate embedding blocks produced by other workers as reusable (and potentially stale) results. Accordingly, the embedding matrix $\mathbf{H}^{(\ell)}$ in Eq. (2.3) is composed of two categories of blocks: (i) freshly computed submatrices from ACTIVE workers that have just broadcast their updates, and (ii) cached historical submatrices from STALE workers whose embedding changes are small enough to ignore at this iteration. Each GPU maintains a cache that stores these historical embeddings, retaining only those considered sufficiently fresh.

Let [ ] denote vertical concatenation of matrix blocks. We use the processor states ACTIVE and STALE to indicate whether process $P(i)$ broadcasts the newest embedding submatrix $\mathbf{H}_i^{(\ell)}$ or, instead, relies on its cached counterpart $\tilde{\mathbf{H}}_i^{(\ell)}$:

$$
\begin{aligned}
\mathbf{H}^{(\ell)} &= \sigma\left(\left[\mathbf{H}_i^{(\ell-1)}\right]_{i=1}^{P}, \hat{\mathbf{A}}; \mathbf{W}^{(\ell-1)}\right) \\
&\approx \sigma\left(\left[\mathbf{H}_{i:P(i)\Leftarrow\text{ACTIVE}}^{(\ell-1)} \mid \tilde{\mathbf{H}}_{i:P(i)\Leftarrow\text{STALE}}^{(\ell-1)}\right]_{i=1}^{P}, \hat{\mathbf{A}}; \mathbf{W}^{(\ell-1)}\right).
\end{aligned}
\tag{5.1}
$$

5.2.4 Skip-Broadcast

Under a decentralized architecture, the key challenge is to adapt collective communication (e.g., one-to-all broadcast) so that workers can safely reuse historical embeddings under bounded staleness. This is non-trivial because common decentralized implementations [4] typically follow a bulk-synchronous execution style, which leaves little room for selectively mixing fresh and stale embedding blocks. To address this issue, we introduce a lightweight communication primitive that is straightforward to implement and does not rely on any centralized parameter server. In particular, we design Skip-Broadcast, which enables the communication topology to be reshaped on the fly during training. To enable Skip-Broadcast, Sancus maintains a per-worker status variable Flag(i) on worker i (with $i \in [1, p]$) to track whether the embedding block computed by that worker, $\mathbf{H}_i$, should be broadcast. When Flag(i) is ACTIVE, worker i disseminates its newest embeddings $\mathbf{H}_i$ to all peers, and each recipient caches the received block. When Flag(i) becomes STALE, Sancus suppresses the broadcast of $\mathbf{H}_i$ for that round, and other workers instead reuse their cached (stale) copy of the corresponding historical embeddings.

Fig. 5.6 Skip-broadcast example, with STALE worker marked red

Figure 5.6 illustrates this mechanism. Suppose GPU 2 is marked STALE via Flag(2); then all other workers fall back to their cached historical block for $\mathbf{H}_2$. Meanwhile, the ring-based pipelined communication schedule is reconfigured to bypass GPU 2 as a sender, directly linking its neighboring GPUs so the pipeline remains continuous. Importantly, GPU 2 is still kept in the broadcast graph as a receiver, allowing it to obtain up-to-date embeddings and gradients from others. In effect, whenever the embedding portion produced by a worker remains within the bounded-staleness constraint, Skip-Broadcast replaces the original broadcast for that worker. By avoiding transmissions from STALE workers, Sancus further lowers communication overhead. The status Flag(i) is examined at every iteration and updated as needed, enabling continual, seamless reshaping of the ring-based communication topology.

5.2.5 *Bounded Embedding Staleness*

Once Skip-Broadcast is enabled, different workers may consume embedding blocks produced in different iterations, yielding a mixed-version embedding view across the system. To keep this system staleness under control, Sancus enforces bounded embedding staleness. Prior studies on bounded-gradient staleness [16, 19, 20] mainly target stochastic gradient descent (SGD): the goal is to preserve convergence by limiting the harm introduced by outdated gradients. Our setting is fundamentally different. Here, stale embeddings are not merely tolerated as an artifact of asynchrony; instead, we intentionally reuse them to reduce communication. To make this safe, we introduce a set of novel bounded embedding staleness metrics ϵ that explicitly quantify and constrain the approximation error induced by stale embeddings.

Bounded-Staleness Definition Variants. We next formalize three notions of historical-embedding staleness, including a definition based on the embedding-variation gap. We begin with an epoch-based, fixed tolerance that bounds the number of outdated epochs allowed for each local update on processor $P(i)$:

Definition 5.2 (*Epoch-Fixed Embedding Staleness*) Consider all processors in a decentralized GNN. Let $\tilde{e}$ be the epoch index at which a stale intermediate embedding was produced, and let e be the current epoch. We define a fixed-epoch-staleness budget ϵ_E as the maximum number of stale epochs that can be tolerated, i.e., $|\tilde{e} - e| \leq \epsilon_E$. Under this rule, intermediate embeddings used during model computation are broadcast only once every ϵ_E epochs.

Nevertheless, the epoch-fixed rule can still make the system behave largely synchronously. To better fit a decentralized setting, especially when workers progress at different speeds, we further introduce two more flexible staleness metrics that allow a worker to reuse embeddings from adaptively selected past iterations. These metrics enable finer-grained control of staleness under heterogeneous execution rates.

Definition 5.3 (*Epoch-Adaptive Embedding Staleness*) For each processor $P(i)$ in decentralized GNN training, we define the epoch-adaptive embedding staleness as the maximum allowable epoch gap ϵ_A between the embeddings used by $P(i)$ and those produced by all of its in-coming neighbor processors. Specifically, $P(i)$ is required to broadcast its latest results after it has received (and potentially reused) embeddings from each in-coming neighbor that are at most ϵ_A epochs old.

Definition 5.4 (*Epoch-Adaptive Variation-Gap Embedding Staleness*) For each processor $P(i)$ in decentralized GNN training, we define a value-based tolerance on stale embeddings by bounding their maximum variation gap. In particular, the epoch-adaptive variation-gap embedding staleness is given by $\|\tilde{\mathbf{H}}_i^{(\ell)} - \mathbf{H} * i^{(\ell)}\| \leq \epsilon * H$, where ϵ_H is the largest embedding-change magnitude that can be tolerated. Under this criterion, intermediate embeddings are broadcast adaptively whenever the variation gap exceeds ϵ_H, independent of how many epochs have been skipped.

Because the above metrics are evaluated locally on each worker, Sancus requires neither a centralized controller nor a global monitor, and thus preserves the decentralized design. Moreover, these general definitions can be readily instantiated in specific distributed GNN systems, serving as practical metrics for studying how stale intermediate results influence distributed training.

Bounded-Staleness-Check Procedure. We now detail how Sancus tests cached historical embeddings against the staleness bounds in Eq. (5.2.5), which is essential for handling the mixed-version embedding view that arises under Skip-Broadcast. The key conceptual shift from conventional distributed ML is that staleness constraints are imposed on intermediate embeddings $\mathbf{H}_i$ rather than on gradients. This choice aligns with our objective: we purposely reuse stale embeddings to cut down avoidable communication in distributed GNN training. Yet, stale embeddings introduce approximation error, so we must bound their impact. In a decentralized environment, this calls for an efficient per-worker mechanism. Accordingly, Sancus equips every worker with a lightweight local state tracker that supports fast staleness checks without requiring any global coordination.

We draw inspiration from the database concept of version control [21], but implement it in a fully decentralized manner to track freshness. Let Ver denote the current epoch index. On worker i, whenever the newest embedding block $\mathbf{H}_j$ is received from worker j, worker i attaches a local version tag $Ver_i(j)$, which records the epoch

Algorithm 2 Embedding staleness check based on Definition 5.2

Input: Current epoch number Ver; Cached embedding version $Ver(j)$

procedure STALE( )
    if $Ver - Ver(j) > \epsilon_E$ **then** ▷ Bounded staleness exceeded
        $F(i) \leftarrow ACTIVE$ ▷ Set flag ACTIVE to indicate too-stale state
    else $F(i) \leftarrow STALE$ ▷ Staleness within tolerance
    end if
end procedure

Algorithm 3 Embedding staleness check based on Definition 5.3

Input: Current epoch number Ver; Cached embedding version $Ver_i(j)$

1: **procedure** STALE( )
2:     **for all** process $P(i)$ in parallel **do**
3:         **for** $j = 1$ to p **do**
4:             ▷ Worker i exceeds staleness bound on cached $\tilde{\mathbf{H}}_j^{\ell-1}$
5:             **if** $Ver - Ver_i(j) > \epsilon_A$ **then**
6:                 $F(j) \leftarrow ACTIVE$ ▷ Indicate too-stale state for $\tilde{\mathbf{H}}_j^{\ell-1}$
7:                 EndProcedure for $P(j)$
8:             **end if**
9:         **end for**
10:     **end for**
11: **end procedure**

at which $\mathbf{H}_j$ was produced. Worker i maintains such version tags for all received blocks $\mathbf{H}_j$ with $j \in [1, \ldots, p]$.

We start with the checking procedure for the epoch-fixed embedding staleness in Definition 5.2. To support this rule, Algorithm 1 requires a minimal change at Line 8: instead of caching only the embedding block, i.e., CACHE$\mathbf{H}_j^{(\ell-1)}$, the worker also stores its associated epoch tag, i.e., CACHE$\mathbf{H}_j^{(\ell-1)}$, $Ver(j)$. Here $Ver(j)$ records the version (epoch index) of the most recently broadcast $\mathbf{H}_j^{(\ell-1)}$. After completing each forward and backward pass, training advances to the next epoch. Algorithm 2 summarizes how the epoch-fixed staleness check is performed. We omit the worker subscript in $Ver_i(j)$ because this basic bound is identical across all workers. As an illustration, Fig. 5.7a shows the case $\epsilon_{\mathrm{E}} = 1$, where embedding broadcasts are skipped every other epoch.

We next consider epoch-adaptive embedding staleness (Definition 5.3). Supporting this criterion again requires augmenting the caching operation at Line 8 of Algorithm 1, but now the version tag is maintained per receiver. Concretely, CACHE$\mathbf{H}_j^{(\ell-1)}$ is replaced by CACHE$\mathbf{H}_j^{(\ell-1)}$, $Ver_i(j)$, where $Ver_i(j)$ records, on processor i, the epoch at which the most recently received $\mathbf{H}_j^{(\ell-1)}$ from worker j was computed. Algorithm 3 presents the corresponding staleness-check routine. If worker i finds that its cached block $\tilde{\mathbf{H}} * j^{\ell-1}$ for some worker j is older than ϵ_{A} epochs (Line 5), then i is effectively advancing faster than j and the cached $\tilde{\mathbf{H}} * j$ is deemed overly stale. In that case, $F(j)$ is set to ACTIVE, prompting worker j to broadcast an updated $\mathbf{H} * j$ soon. This rule

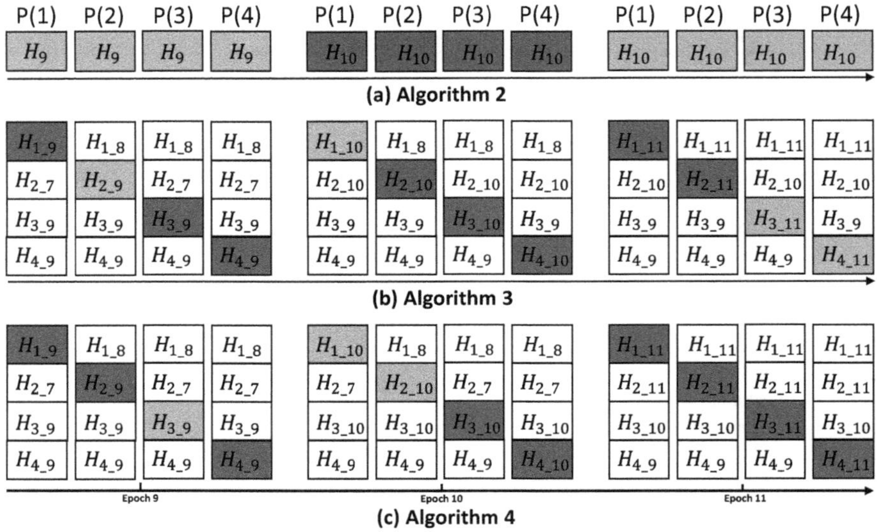

Fig. 5.7 Example training behavior under each bounded embedding staleness metric. Worker i is denoted by $P(i)$ for $i \in [1, 4]$, and H_{i_e} represents the embedding results produced by process i at epoch e. Workers marked ACTIVE are shown in green, while STALE workers are shown in red. In the examples corresponding to Algorithms 2 and 3, we set $\epsilon_E = \epsilon_A = 1$. For Algorithm 4, we provide a separate toy instance for illustration

Algorithm 4 Embedding staleness check based on Definition 5.4

Input: Current $\mathbf{H}_i^\ell$ computed in $P(i)$; Cached embeddings $\tilde{\mathbf{H}}_i^\ell$

```
procedure STALE( )
    if ||H_i^ℓ − H̃_i^ℓ|| > ε_H then          ▷ Bounded staleness exceeded
        F(i) ← ACTIVE          ▷ Set flag ACTIVE to indicate too-stale state
    else F(i) ← STALE                     ▷ Staleness within tolerance
    end if
end procedure
```

enforces that a worker proceeds with broadcasting its newest results only after it has received sufficiently recent updates from every other worker (i.e., no older than ϵ_A epochs). Figure 5.7b illustrates the behavior for $\epsilon_A = 1$. At Epoch 9, worker 2's cached result $\mathbf{H} * 2\_7$ becomes unacceptable because $9 - 7 = 2 > \epsilon_A = 1$, so worker 2 is switched to ACTIVE and broadcasts refreshed embeddings in Epoch 10. Subsequently, worker 1 is marked ACTIVE because other workers are still using $\mathbf{H} * 1\_8$, which now violates the bound. Similarly, workers 3 and 4 are designated as ACTIVE in Epoch 11.

Finally, we consider the epoch-adaptive variation-gap embedding staleness in Definition 5.4. Under this metric, staleness is decided solely by how much a worker's embeddings change, and the decision is made locally on each worker. Algorithm 4 details the corresponding checking procedure. If the change in $\mathbf{H}_i^\ell$ stays within the prescribed tolerance, then broadcasting the newest block is unnecessary: the system can continue to reuse the cached historical embedding $\tilde{\mathbf{H}}_i^\ell$ at other workers. Once the variation exceeds the threshold, however, the current $\mathbf{H}_i^\ell$ must be broadcast and the cache refreshed accordingly. Notably, this criterion does not require any explicit version tracking. As illustrated in Fig. 5.7c, because the decision depends only on the embedding-variation gap, a worker may remain STALE for an adaptive number of epochs before being forced to broadcast again.

5.3 Theoretical Analysis

In this section, we develop the theoretical foundations of Sancus. We first derive an upper bound on communication overhead using the α–β communication-cost model in Sect. 5.3.1. We then analyze the approximation induced by stale embeddings by bounding the errors of both embeddings and gradients, and establish convergence guarantees for Sancus in Sect. 5.3.2.

5.3.1 Communication Cost Analysis

The amount of communication that can be eliminated is governed by the staleness tolerance introduced in Sect. 5.2.5. Intuitively, allowing a wider tolerance permits more broadcasts to be skipped, and therefore yields a larger reduction in communication overhead. To quantify this effect, we derive communication bounds under staleness by first establishing a baseline bound without staleness. Our derivation follows the analyses in Tripathy et al. [4] and Chan et al. [22], adopting the standard α–β cost model [4, 22]. In this model, α captures a fixed startup latency that is independent of message size, while β represents the per-word transfer time (i.e., the inverse of bandwidth) to deliver data to the destination worker. Under this formulation, broadcasting a single message of F words to p processes has a communication lower bound of $\mathcal{O}(\alpha + F\beta)$ [4], where F is the embedding dimension. Likewise, the all-reduce cost COSTar in Eq. (2.6) admits a lower bound of $\mathcal{O}\left(\log p; \alpha + 2\frac{p-1}{p}F\beta\right)$ [22].

Consequently, when no staleness is permitted, each process in the forward pass of Eq. (2.3) must broadcast its intermediate embedding blocks $\mathbf{H}_j^{(\ell)}$ to the other $(p-1)$ processes, leading to a per-process communication lower bound of $(p-1)\left(\alpha + \frac{N}{p}F\beta\right)$. In backpropagation, the corresponding lower bound is $(p-1)\left(\alpha + \frac{N}{p}F\beta\right)$ for Eq. (2.6) and $\left(\log p; \alpha + 2\frac{p-1}{p}F^2\beta\right)$ for Eq. (2.6), which performs AllReduce over $(F \times F)$-matrices.

Hence, the total communication COST$comm$ has a lower bound:

$$\mathcal{O}\left(L\left((2(p-1)+\log p)\alpha+2\frac{p-1}{p}F(N+F)\beta\right)\right). \tag{5.2}$$

Let e_{T} denote the total number of epochs, and let e_{R} be the number of epochs in which Skip-Broadcast is triggered so that historical embeddings are reused. Under these definitions, we can derive a lower bound on the overall communication cost, denoted by COST$comm^{\star}$, for our staleness-aware parallel algorithm. The total cost naturally separates into two components:

(1) during epochs that transmit fresh results, communication follows the baseline behavior, and the cost matches COST$comm$ in Eq. (5.2);
(2) during epochs that reuse historical embeddings, the broadcast component is eliminated, so the remaining communication is dominated by the AllReduce in Eq. (2.6) over COSTar.

Accordingly, the communication-cost bound can be written as:

$$\begin{aligned}\mathrm{COST}comm^{*} &= (e_{\mathrm{T}}-e_{\mathrm{R}})\,\mathrm{COST}comm+e_{\mathrm{R}}L\mathrm{COST}ar\\ &= L\left\{\left[2(p-1)(e_{\mathrm{T}}-e_{\mathrm{R}})+e_{\mathrm{T}}\log p\right]\alpha\right.\\ &\quad\left.+2\,(p-1)/p\;F\left[N(e_{\mathrm{T}}-e_{\mathrm{R}})+e_{\mathrm{T}}F\right]\beta\right\}.\end{aligned}$$

Therefore, the resulting expression yields a tighter communication bound, although the exact relative savings are not immediately transparent from the bound itself. As we show empirically in Sect. (5.4.2), our framework can reduce communication cost by up to 74%.

5.3.2 Convergence Analysis

In this section, building on bounded embedding staleness, we establish theoretical guarantees for Sancus, including bounds on the approximation errors of intermediate embedding results and gradients, together with convergence assurances. In particular, we show that caching and repeatedly reusing historical information to reduce communication introduces approximation errors that remain controllable under our staleness constraints. Concretely, Sect. 5.3.2 derives the convergence guarantee through the following steps:

- Proposition 5.1 provides the basic inequality tools and algebraic facts needed for the analysis;
- Lemma 5.1 shows that, under bounded embedding staleness, the approximated intermediate matrix results stay close to their exact counterparts throughout training;
- Lemma 5.2 further establishes that the resulting gradient approximations remain close to the exact gradients;

- Theorem 5.1 concludes that model weights evolve slowly enough that gradients are asymptotically unbiased, which in turn guarantees convergence.

Due to space constraints, we present the analysis only for the stricter condition in Definition 5.4. The theoretical results for the basic fixed-epoch staleness follow as a corollary by setting $\epsilon_{\mathrm{H}} = \max \Delta_{|\mathbf{H}|} \times \epsilon_{\mathrm{E}}$, where $\max \Delta_{|\mathbf{H}|}$ denotes the maximum per-epoch variation magnitude of $\mathbf{H}$.

Details. Our proof of Theorem 5.1 follows the overall proof strategy in ([5], Theorem 2), but it is not a direct reuse. In particular, under the assumption that the activation $\sigma(\cdot)$ and the loss gradients $\nabla\mathcal{L}$ are ρ-Lipschitz continuous (which is important given their potentially strong non-linearity), our analysis differs from [5] in three essential ways:

(1) we study full-batch GNN training on the entire graph, i.e., without sampling;
(2) training is carried out in a decentralized parallel setting rather than a centralized scheme;
(3) historical embeddings are reused adaptively, where the number of skipped (stale) epochs is determined automatically, instead of restricting stale activations to be only one epoch old as in many prior approaches.

First of all, we extract the proposition ([5], Proposition B) below as it is needed in our proof:

Proposition 5.1 *Denote* $\|\mathbf{A}\|_\infty = \max_{i,j} |\mathbf{A}(i,j)|$ *and* $col(\mathbf{A})$ *as the column number of matrix* $\mathbf{A}$*, we have* $\|\mathbf{A}+\mathbf{B}\|_\infty \le \|\mathbf{A}\|_\infty + \|\mathbf{B}\|_\infty$, $\|\mathbf{A}\odot\mathbf{B}\|_\infty \le \|\mathbf{A}\|_\infty\|\mathbf{B}\|_\infty$, *and* $\|\mathbf{A}\mathbf{B}\|_\infty \le col(\mathbf{A})\|\mathbf{A}\|_\infty\|\mathbf{B}\|_\infty$.

The proof ([5], Appendix C) is omitted. We further denote C as the maximum number of columns that exists in our proof:

$$C := \max\{col(\hat{\mathbf{A}}), col(\mathbf{H}^{(0)}), \dots, \\ col(\mathbf{H}^{(L)}), col(\mathbf{W}^{(0)}), \dots, col(\mathbf{W}^{(L)})\}.$$

Next, we show that imposing bounded embedding staleness yields a provable bound on the approximation error of intermediate results in decentralized training. This provides theoretical justification for Skip-Broadcast and embedding reuse, demonstrating that communication can be reduced without incurring excessive loss in accuracy.

Lemma 5.1 *Consider a decentralized setting with processes* $P(i)$, $i \in \{1, \dots, p\}$, *where each process uses* N/p *input activations* $\tilde{\mathbf{H}}_i^{(\ell)}$ *at layer* ℓ*. Assume that:*

(1) the activation function $\sigma(\cdot)$ *is* ρ*-Lipschitz continuous;*
(2) the matrices $\hat{\mathbf{A}}_i$, $\tilde{\mathbf{H}}_i$, $\mathbf{H}_i$, *and* $\mathbf{W}_i$ *have bounded entries, i.e., are bounded by a constant* B*;*
(3) historical embeddings $\tilde{\mathbf{H}}_i^{(\ell)}$ *approximate the exact embeddings* $\mathbf{H}_i^{(\ell)}$ *within the staleness budget* ϵ_H*, namely* $\|\tilde{\mathbf{H}}_i^{(\ell)} - \mathbf{H} * i^{(\ell)}\| * \infty \le \epsilon_H$.

Then, the approximation errors of the intermediate outputs $\tilde{\mathbf{T}}_i$ and $\tilde{\mathbf{Z}}_i$ are also bounded:

(1) $\|\tilde{\mathbf{T}}_i^{(\ell)} - \mathbf{T} * i^{(\ell)}\| * \infty \leq pCB\epsilon_H$;
(2) $\|\tilde{\mathbf{Z}}_i^{(\ell)} - \mathbf{Z} * i^{(\ell)}\| * \infty \leq C^2B^2\epsilon_H$.

Here $\ell \in [1, L]$ denotes the ℓth layer.

Proof In our decentralized GNN training, the block $\tilde{\mathbf{H}}_i$ used by a worker is taken to be either the current (up-to-date) embeddings $\mathbf{H}_i$ or a reused historical snapshot from an earlier epoch. In the latter case, $\tilde{\mathbf{H}}_i = \mathbf{H}_s$ for some stale epoch $s < i$, and the reuse is permitted only when the deviation is bounded, i.e., $\|\mathbf{H}_i - \mathbf{H}_s\| \leq \epsilon_{\mathrm{H}}$.

By Proposition 5.1, we have

$$\left\|\tilde{\mathbf{T}}_i^{(\ell)} - \mathbf{T}_i^{(\ell)}\right\|_\infty = \left\|\sum_{j=1}^{p} \hat{\mathbf{A}}_{ij}\tilde{\mathbf{H}}_j^{(\ell)} - \sum_{j=1}^{p} \hat{\mathbf{A}}_{ij}\mathbf{H}_j^{(\ell)}\right\|_\infty$$
$$\leq p \max_{j=1}^{p} \left\|\hat{\mathbf{A}}_{ij}\left(\tilde{\mathbf{H}}_j^{(\ell)} - \mathbf{H}_j^{(\ell)}\right)\right\|_\infty \leq pCB\epsilon_{\mathrm{H}},$$

and also

$$\left\|\tilde{\mathbf{Z}}_i^{(\ell)} - \mathbf{Z}_i^{(\ell)}\right\|_\infty = \left\|\hat{\mathbf{A}}_i\tilde{\mathbf{H}}_i^{(\ell-1)}\mathbf{W}_i^{(\ell-1)} - \hat{\mathbf{A}}_i\mathbf{H}_i^{(\ell-1)}\mathbf{W}_i^{(\ell-1)}\right\|_\infty$$
$$\leq C^2\left\|\hat{\mathbf{A}}_i\right\|_\infty \left\|\tilde{\mathbf{H}}_i^{(\ell-1)} - \mathbf{H}_i^{(\ell-1)}\right\|_\infty \left\|\mathbf{W}_i^{(\ell-1)}\right\|_\infty \leq C^2B^2\epsilon_{\mathrm{H}}.$$

Then, we prove that the approximate gradients $\tilde{\delta}^{(\ell)} = \nabla_{\tilde{\mathbf{Z}}^{(\ell)}}\tilde{\mathcal{L}}$ are close to the exact gradients $\delta_i^{(\ell)} = \nabla_{\mathbf{Z}_i^{(\ell)}}\mathcal{L}$.

Lemma 5.2 *For any process $P(i)$ with $i \in [1, p]$, suppose that:*

(1) *the activation $\sigma(\cdot)$ and the loss gradient $\nabla\mathcal{L}$ are ρ-Lipschitz continuous;*
(2) *the quantities $\hat{\mathbf{A}}_i$, $\mathbf{W} * i$, $\delta^{(\ell)}$, and $\sigma'(\mathbf{Z} * i)$ are uniformly bounded by a constant B.*

Then the gradient errors induced by stale intermediate states remain bounded. In particular, the approximation errors of $\nabla * \tilde{\mathbf{Z}} * i^{(\ell)}\tilde{\mathcal{L}}$ and $\nabla * \mathbf{W} * i^{(\ell)}\tilde{\mathcal{L}}$ satisfy:

(1) $\|\nabla * \tilde{\mathbf{Z}} * i^{(\ell)}\tilde{\mathcal{L}} - \nabla * \mathbf{Z} * i^{(\ell)}\mathcal{L}\| * \infty \leq K$;
(2) $\|\nabla * \mathbf{W} * i^{(\ell)}\tilde{\mathcal{L}} - \nabla * \mathbf{W} * i^{(\ell)}\mathcal{L}\| * \infty \leq K$.

Here $\ell \in \{1, \ldots, L\}$ denotes the layer index, and K is a constant determined by ρ, C, B, and ϵ_H.

Proof We bound the approximation error of $\nabla_{\tilde{\mathbf{Z}}^{(\ell)}}\tilde{\mathcal{L}}$ via induction over layers. The argument relies on the ρ-Lipschitz continuity of the activation $\sigma(\cdot)$ and of $\nabla_{\mathbf{Z}^{(\ell)}}\mathcal{L}$, together with the backpropagation recurrence in Eq. (2.6), the inequality tools in Proposition 5.1, and the intermediate-error bounds from Lemma 5.1.

Base case. Consider the last layer L. Using the ρ-Lipschitz property of $\nabla_{\mathbf{Z}^{(\ell)}}\mathcal{L}$ and the bound in Lemma 5.1, we obtain

$$\left\| \nabla_{\tilde{\mathbf{Z}}_i^{(L)}} \tilde{\mathcal{L}} - \nabla_{\tilde{\mathbf{Z}}_i^{(L)}} \mathcal{L} \right\|_\infty \le \rho \left\| \tilde{\mathbf{Z}}_i^{(L)} - \mathbf{Z}_i^{(L)} \right\|_\infty \le \rho C^2 B^2 \epsilon_{\mathrm{H}}.$$

So the statement holds for $\ell = L$, where $K_{(L)} = \rho C^2 B^2 \epsilon_{\mathrm{H}}$.
Induction hypothesis. Suppose that $\|\nabla_{\tilde{\mathbf{Z}}_i^{(\ell')}} \tilde{\mathcal{L}} - \nabla_{\tilde{\mathbf{Z}}_i^{(\ell')}} \mathcal{L}\|_\infty \le K_{(\ell')}$ holds for all $\ell' > \ell$. For layer ℓ, by Eq. (2.6), Proposition 5.1, and Lemma 5.1, we have

$$\begin{aligned}
\left\| \nabla_{\tilde{\mathbf{Z}}_i^{(\ell)}} \tilde{\mathcal{L}} - \nabla_{\tilde{\mathbf{Z}}_i^{(\ell)}} \mathcal{L} \right\|_\infty &= \| \tilde{\delta}_i^{(\ell+1)} \hat{\mathbf{A}}_i \left(\mathbf{W}_i^{(\ell)}\right)^\mathsf{T} \odot \sigma'\left(\tilde{\mathbf{Z}}_i^{(\ell)}\right) \\
&\quad - \delta_i^{(\ell+1)} \hat{\mathbf{A}}_i \left(\mathbf{W}_i^{(\ell)}\right)^\mathsf{T} \odot \sigma'\left(\mathbf{Z}_i^{(\ell)}\right) \|_\infty \\
&\le C^2 \{ \left\| \tilde{\delta}_i^{(\ell+1)} \right\|_\infty \left\| \hat{\mathbf{A}}_i \right\|_\infty \left\| \left(\mathbf{W}_i^{(\ell)}\right)^\mathsf{T} \right\|_\infty \left\| \sigma'\left(\tilde{\mathbf{Z}}_i^{(\ell)}\right) - \sigma'\left(\mathbf{Z}_i^{(\ell)}\right) \right\|_\infty \\
&\quad + \left\| \tilde{\delta}_i^{(\ell+1)} - \delta_i^{(\ell+1)} \right\|_\infty \left\| \hat{\mathbf{A}}_i \right\|_\infty \left\| \left(\mathbf{W}_i^{(\ell)}\right)^\mathsf{T} \right\|_\infty \left| \sigma'\left(\mathbf{Z}_i^{(\ell)}\right) \right|_\infty \} \\
&\le C^2 \left(B^3 \rho (C^2 B^2 \epsilon_{\mathrm{H}}) + K_{(\ell+1)} B^3 \right) = C^2 B^3 \left(K_{(L)} + K_{(\ell+1)} \right).
\end{aligned}$$

Setting $K_{(\ell)} = C^2 B^3 \left(K_{(L)} + K_{(\ell+1)}\right)$, the inequality can also be kept for layer ℓ, which completes the inductive step.
Conclusion. This completes the induction and thus establishes the desired bound.

A corresponding bound for the approximation error of $\nabla_{\mathbf{W}_i^{(\ell)}} \tilde{\mathcal{L}}$ follows from an analogous argument; we omit the details due to space constraints.

Finally, we establish a convergence guarantee for the proposed decentralized GNN training scheme under bounded embedding staleness.

Theorem 5.1 *Consider an L-layer GNN model with local minimizer* $\mathbf{W}^\star$*, initialization* $\mathbf{W} * (1)$*, and staleness budget* ϵ_H*. Assume that*

(1) the activation $\sigma(\cdot)$ *and the loss gradient* $\nabla\mathcal{L}$ *are* ρ*-Lipschitz continuous;*
(2) for $\hat{\mathbf{A}}$*,* $\mathbf{H}$*, and* $\mathbf{W}$*, the gradients with respect to the weights are uniformly bounded, i.e.,* $\|\nabla * \mathbf{W}\tilde{\mathcal{L}}\|_\infty$*,* $\|\nabla_{\mathbf{W}}\mathcal{L}\|_\infty$*, and* $\|\nabla\mathcal{L}(\mathbf{W})\|_\infty$ *are all bounded by a constant* $G > 0$*;*
(3) the loss function $\mathcal{L}(\mathbf{W})$ *is* ρ*-smooth.*

Then there exists a constant $K > 0$ *such that, for any* $N > L\epsilon_H$*, if the distributed GNN is trained in parallel with bounded staleness for at most N iterations and an iterate index* $R \in [1, \ldots, N]$ *is chosen uniformly at random, using learning rate* $\eta = \min\{\frac{1}{\rho}, \frac{1}{\sqrt{N}}\}$*, we have*

$$\mathbb{E}_R \|\nabla\mathcal{L}(\mathbf{W}_{(R)})\|_F^2 \le 2\frac{\mathcal{L}(\mathbf{W}_{(1)}) - \mathcal{L}(\mathbf{W}^\star) + \frac{\rho K}{2}}{\sqrt{N}}, \tag{5.3}$$

where the learning rate $\eta = \min\{\frac{1}{\rho}, \frac{1}{\sqrt{N}}\}$*.*

Proof For convenience, let $\Delta_{(i)} = \nabla_{\mathbf{W}_{(i)}} \tilde{\mathcal{L}} - \nabla\mathcal{L}(\mathbf{W}_{(i)})$, by Lemma 5.2 and the ρ-smoothness of the $\mathcal{L}(\mathbf{W})$, we derive $\mathcal{L}(\mathbf{W}_{(i+1)})$

$$
\begin{aligned}
&\leq \mathcal{L}(\mathbf{W}_{(i)}) + \langle \nabla \mathcal{L}(\mathbf{W}_{(i)}), \mathbf{W}_{(i+1)} - \mathbf{W}_{(i)} \rangle + \frac{\rho}{2}\eta^2 \|\nabla_{\mathbf{W}_{(i)}} \tilde{\mathcal{L}}\|_F^2 \\
&= \mathcal{L}(\mathbf{W}_{(i)}) - \eta \langle \nabla \mathcal{L}(\mathbf{W}_{(i)}), \nabla_{\mathbf{W}_{(i)}} \tilde{\mathcal{L}} \rangle + \frac{\rho}{2}\eta^2 \|\nabla_{\mathbf{W}_{(i)}} \tilde{\mathcal{L}}\|_F^2 \\
&= \mathcal{L}(\mathbf{W}_{(i)}) - \eta \langle \nabla \mathcal{L}(\mathbf{W}_{(i)}), \Delta_{(i)} \rangle - \eta \|\nabla \mathcal{L}(\mathbf{W}_{(i)})\|_F^2 \\
&\quad + \frac{\rho}{2}\eta^2 \left(\|\Delta_{(i)}\|_F^2 + \|\nabla \mathcal{L}(\mathbf{W}_{(i)})\|_F^2 + 2\langle \Delta_{(i)}, \nabla \mathcal{L}(\mathbf{W}_{(i)}) \rangle \right) \\
&\leq \mathcal{L}(\mathbf{W}_{(i)}) - (\eta - \frac{\rho}{2}\eta^2) \|\nabla \mathcal{L}(\mathbf{W}_{(i)})\|_F^2 + \frac{\rho}{2}\eta^2 \|\Delta_{(i)}\|_F^2.
\end{aligned}
$$

By Lemma 5.2, $\|\Delta_{(i)}\|_F^2 \leq \|\nabla_{\mathbf{W}_{(i)}} \tilde{\mathcal{L}}\|_\infty + \|\nabla \mathcal{L}(\mathbf{W}_{(i)})\|_\infty \leq 2G^2 \leq K$. Thus,

$$
\mathcal{L}(\mathbf{W}_{(i+1)}) \leq \mathcal{L}(\mathbf{W}_{(i)}) - (\eta - \frac{\rho}{2}\eta^2) \|\nabla \mathcal{L}(\mathbf{W}_{(i)})\|_F^2 + \frac{\rho}{2}\eta^2 K. \tag{5.4}
$$

For all i, we sum up Eq. (5.4) and rearrange the terms:

$$
\left(\eta - \frac{\rho}{2}\eta^2\right) \sum_{i=1}^{N} \left\| \nabla \mathcal{L}(\mathbf{W}_{(i)}) \right\|_F^2 \leq \mathcal{L}(\mathbf{W}_{(1)}) - \mathcal{L}(\mathbf{W}^\star) + \frac{\rho}{2}\eta^2 K N. \tag{5.5}
$$

Recall $\eta = \min\{\frac{1}{\rho}, \frac{1}{\sqrt{N}}\}$, we divide both sides of Eq. (5.5) by $N(\eta - \frac{\rho}{2}\eta^2)$, then we obtain $\mathbb{E}_R \|\nabla \mathcal{L}(\mathbf{W}_{(R)})\|_F^2$:

$$
\begin{aligned}
&= \frac{1}{N} \sum_{i=1}^{N} \|\nabla \mathcal{L}(\mathbf{W}_{(i)})\|_F^2 \leq 2 \frac{\mathcal{L}(\mathbf{W}_{(1)}) - \mathcal{L}(\mathbf{W}^\star) + \frac{\rho}{2}\eta^2 K N}{N\eta(2 - \rho\eta)} \\
&\leq 2 \frac{\mathcal{L}(\mathbf{W}_{(1)}) - \mathcal{L}(\mathbf{W}^\star)}{N\eta} + \rho\eta K \leq 2 \frac{\mathcal{L}(\mathbf{W}_{(1)}) - \mathcal{L}(\mathbf{W}^\star) + \frac{\rho K}{2}}{\sqrt{N}}.
\end{aligned}
$$

In particular, $\mathbb{E}_R \|\nabla \mathcal{L}(\mathbf{W}_{(R)})\|_F^2 \to 0$ when $N \to \infty$. The above concludes that the convergence is guaranteed.

5.4 Experiments

We conduct an empirical evaluation of Sancus with the goal of answering four key questions.

- How effectively does Sancus reduce communication across different system configurations, and how does it compare against SOTA methods? (Sects. 5.4.2 and 5.4.3)
- Can Sancus lower communication while preserving model accuracy through staleness-aware reuse? (Sect. 5.4.4)
- What does the skip-epoch pattern look like in practice, and does it adapt in a way that manages staleness effectively? (Sect. 5.4.4)
- What memory overhead is introduced by caching historical embeddings? (Sect. 5.4.4)

Table 5.2 Summary of the graph data statistics used in our experiments to evaluate the proposed framework ("*m*": multi-class classification)

Dataset	$\lvert \mathcal{V} \rvert$	$\lvert \mathcal{E} \rvert$	$\lvert F \rvert$	# Class	Byte size*
Flickr	89,250	899,756	500	7	529 MB
Reddit	232,965	11,606,919	602	41	3.53 GB
Amazon	1,598,960	132,169,734	200	107 (m)	2.34 GB
ogbn-products	2,449,029	61,859,140	100	47	1.38 GB
ogbn-papers100M	111,059,956	1,615,685,872	200	172	56.2 GB

5.4.1 Datasets and Implementations

We evaluate Sancus on five widely used [23] large-scale benchmark datasets [24, 25], summarized in Table 5.2. Flickr and Reddit are evaluated with single-class node classification, whereas Amazon, ogbn-products, and ogbn-papers100M correspond to multi-class classification tasks. Concretely, Flickr captures relations among images that share common properties. The Reddit graph contains posts and user comments, and the goal is to predict the topical community of each post. For Amazon and ogbn-products, nodes represent products and edges connect products co-purchased by the same customer; the task is to assign multiple labels to product nodes. Ogbn-papers100M is a citation network where we predict the subject areas of papers. All datasets use the fixed-partition splits [24, 25].

We build Sancus on top of a PyTorch [26] implementation of classical decentralized parallel algorithms [15] that have been adapted for distributed GNN training [4]. Sancus departs from this baseline in a fundamental way: it reduces communication adaptively by skip-broadcasting cached historical embeddings according to our bounded embedding staleness criteria. Concretely, we implement the fixed-epoch bound ϵ_{E} in Definition 5.2 using Algorithm 1 together with Algorithm 2, and use it as a baseline variant denoted SCS-E. This variant automatically skips broadcasts and reuses cached stale results, triggering a broadcast once every ϵ_{E} epochs. We further implement the epoch-adaptive bound ϵ_{A} in Definition 5.3 via Algorithms 1 and 3, denoted SCS-A. Finally, we implement the variation-gap bound ϵ_{H} in Definition 5.4 with Algorithms 1 and 4, denoted SCS-H; this variant can skip-broadcast historical embeddings for an adaptively determined number of epochs. In addition, we implement SkipG [27], which applies bounded gradient staleness from traditional distributed training, to demonstrate that bounded embedding staleness is more effective in our setting.

In decentralized training, Sancus integrates three essential components: a cache for historical embeddings, a version controller with a bounded-staleness checker, and the Skip-Broadcast operator. At each layer $\ell \in [1, L]$, the staleness checker either (i) measures the difference between the current epoch index and the stored versions of cached blocks or (ii) directly compares the most recent embeddings $\mathbf{H}_i^{\ell-1}$ against their cached historical counterparts to determine whether the configured bound $\epsilon_E/\epsilon_A/\epsilon_H$ has been violated. When a cached block is deemed too stale, the historical-embedding cache is refreshed accordingly, and we only keep the newly received embeddings. Following common practice [5, 18, 28], we include several warm-up epochs in which no broadcasts are skipped, since embedding variations can be large early in training and aggressive reuse may hurt performance. Finally, Sancus is orthogonal to systems that employ arbitrary matrix partitioning, and it is straightforward to deploy and extend due to the broad adoption of PyTorch in the community.

Our experiments consider four GPU configurations: ① eight RTX 2080 Ti GPUs connected by PCIe 3.0 × 16; ② two servers connected via 10Gbps Ethernet, each equipped with four RTX 2080 Ti GPUs over PCIe 3.0; ③ four A100 40GB GPUs interconnected by NVLink; and ④ four V100 32 GB GPUs interconnected by NVLink. We run Sancus on ① for Flickr, Reddit, Amazon, and ogbn-products, and evaluate ogbn-products additionally under ①, ②, and ③. For the largest dataset, ogbn-papers100M, we use ③. We also use ④—a widely adopted setup—for end-to-end comparisons against SOTA systems [2–4, 6, 29]. We use GCN [1] with 3 or 4 layers, and vary the hidden dimension across datasets (i.e., 16 for Flickr and Reddit, 256 for Amazon, 128 for ogbn-products, and {16, 32, 64} for ogb-papers100M). In addition, we test different GPU counts (2/4/8) on Flickr under ①, as well as on ogbn-products under ①/②/③. To further validate generality beyond GCN, we also implement GAT [30].

The total number of training epochs e_T is set to 300/300/400/500/200 for Flickr, Reddit, Amazon, ogbn-products, and ogbn-papers100M, respectively. Following established practice for staleness-tolerant training [5, 18, 28], we include 50 warm-up epochs for every dataset. For staleness control, we sweep $\epsilon_E \in [1, 7]$, $\epsilon_A \in [1, 5]$, and $\epsilon_H \in [0.01, 0.05]$ to regulate embedding-variation amplitude. Without loss of generality, Sects. 5.4.2–5.4.4 report the main results using $\epsilon_E = \{1, 5\}$, $\epsilon_A = \{1, 5\}$, and $\epsilon_H = \{0.01, 0.05\}$, while Sect. 5.4.5 analyzes how these bounds influence staleness-aware training.

5.4.2 The Communication Efficiency Evaluation

We first evaluate how much communication Sancus can eliminate across benchmark datasets. These datasets span increasing orders of magnitude, allowing us to stress-test scalability. In Fig. 5.8, we report configurations whose accuracy loss is within 0.01. Relative to the SOTA decentralized baseline CAGNET and the bounded-gradient approach SkipG, all Sancus variants achieve substantially larger savings, reducing communication by an additional 35% to 74% through bounded embedding staleness. While SkipG (based on traditional gradient staleness) can also reduce communication compared to CAGNET by skipping gradient broadcasts, it does so without staleness-aware safeguards for model quality, and thus trades communication for accuracy. As shown in Fig. 5.10, SkipG can incur accuracy drops of up to 5%, whereas Sancus maintains GNN accuracy across all datasets. Moreover, Fig. 5.8

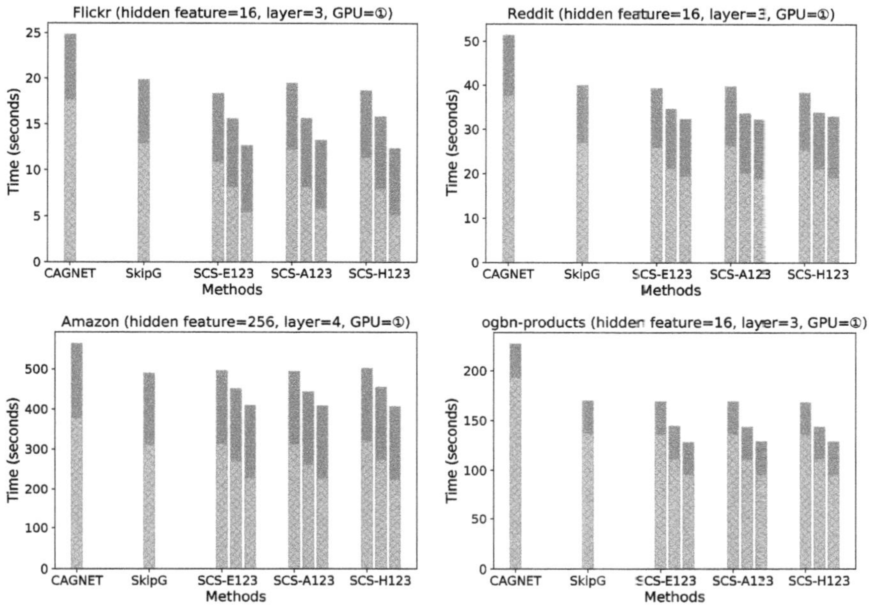

Fig. 5.8 Communication-avoiding performance using all 8 GPUs on Flickr/Reddit/Amazon/ogbn-products datasets. In each subplot, x-axis denotes the methods compared: CAGNET [4], SkipG [27], SCS-E1/E2/E3 with $\epsilon_E = \{1, 2, 3\}$, SCS-A1/A2/A3 with $\epsilon_A = \{1, 2, 3\}$, SCS-H1/H2/H3 with $\epsilon_H = \{0.01, 0.02, 0.03\}$; y-axis denotes the time proportion during training, where blue bar denotes model computation cost and orange bar denotes communication cost

shows that Sancus still exceeds SkipG in communication reduction, delivering an extra 29% to 63% savings depending on the dataset.

To illustrate the strongest outcomes, Sancus reduces communication by about 74% on Flickr using SCS-H3 with $\epsilon_H = 0.03$, by 48% on Reddit using SCS-A3, and by 50% on ogbn-products using SCS-H3. We note that although SCS-E can appear competitive in terms of communication avoidance, its accuracy deteriorates rapidly as ϵ_E increases, as evidenced in Figs. 5.10b, 5.11a, and 5.12b, especially when compared with the more adaptive SCS-A/H variants. In addition, the reported 35% reduction on Amazon and ogbn-papers100M may look modest because both settings intrinsically involve much larger communication volume: Amazon uses a deeper 4-layer GNN with the largest hidden dimension (256) to preserve accuracy, and ogbn-papers100M is the largest dataset in our suite. With stronger accuracy preservation, the adaptive SCS-A/H variants are therefore more robust in practice. Overall, these results confirm the effectiveness of bounded staleness and adaptive communication reduction.

Without loss of generality, we assess the generality of Sancus using the most conservative variants, SCS-E1/A1/H1, which deliver the smallest communication savings. This choice aligns with our goal of isolating how different system config-

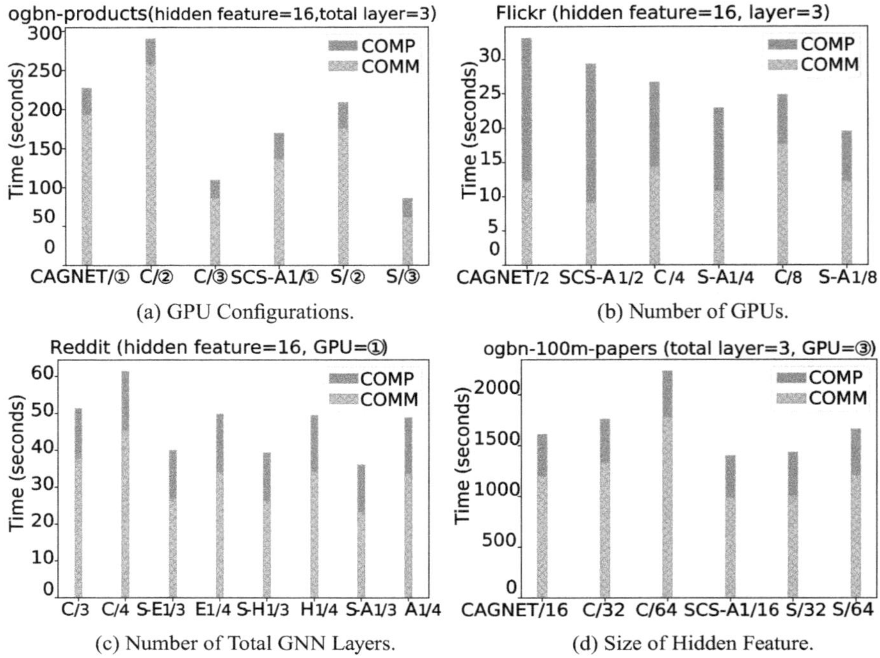

(a) GPU Configurations.

(b) Number of GPUs.

(c) Number of Total GNN Layers.

(d) Size of Hidden Feature.

Fig. 5.9 Performance with system architecture variants (GPU configurations/number of GPUs/GNN layer number/GNN hidden feature size) compared to CAGNET on ogbn-products/Flickr/Reddit/ogbn-papers100M datasets, respectively. In each subplot, y-axis denotes the time proportion during training, where blue bar denotes model computation cost and orange bar denotes communication cost. For the x-axis, we denote the method/GPU configurations in **a**, method/number of GPUs in **b**, method/layer number in **c**, and method/hidden feature size in **d**

urations affect behavior. In Fig. 5.9, we examine architectural variations along four axes (i.e., GPU configurations, GPU count, the total number of GNN layers, and hidden feature size) across ogbn-products, Flickr, Reddit, and ogbn-papers100M. To further quantify end-to-end speedups, we also report an epoch-level time breakdown of computation versus communication in Table 5.3.

As shown in Fig. 5.9a, communication time differs substantially across GPU configurations. Despite this variability, Sancus delivers stable communication reduction under all three settings ①②③, covering both multi-GPU single-node and multi-server deployments, and remaining effective across different GPU types even though the absolute communication cost is strongly hardware dependent. We further report the detailed computation/communication breakdown under ①②③ in Table 5.3. Across these environments, including the multi-server case, SCS-A1 consistently incurs the smallest overall overhead. Figure 5.9b further shows that as the number of GPUs increases, the total training cost relative to CAGNET keeps decreasing. Although raw communication grows with more GPUs, SCS-H substantially mitigates this growth:

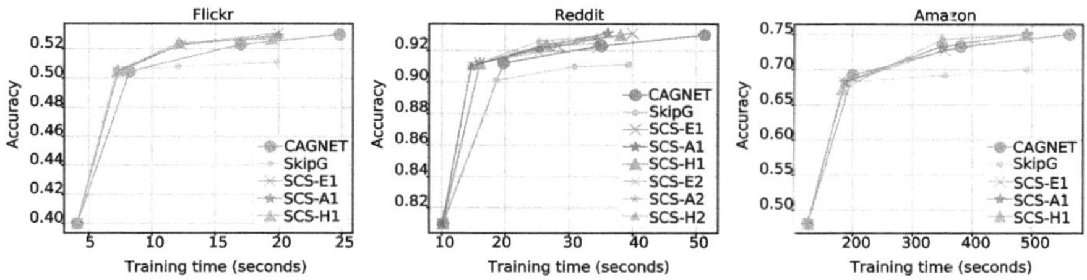

Fig. 5.10 Accuracy results using all eight GPUs on Flickr/Reddit/Amazon datasets, respectively. In each subplot, x-axis denotes the methods compared: CAGNET [4], SkipG [27]. SCS-E1/E2 denotes $\epsilon_E = \{1, 2\}$, SCS-A1/A2 $\epsilon_A = \{1, 2\}$, SCS-H1/H2 $\epsilon_H = \{0.01, 0.02\}$; y-axis denotes the time proportion

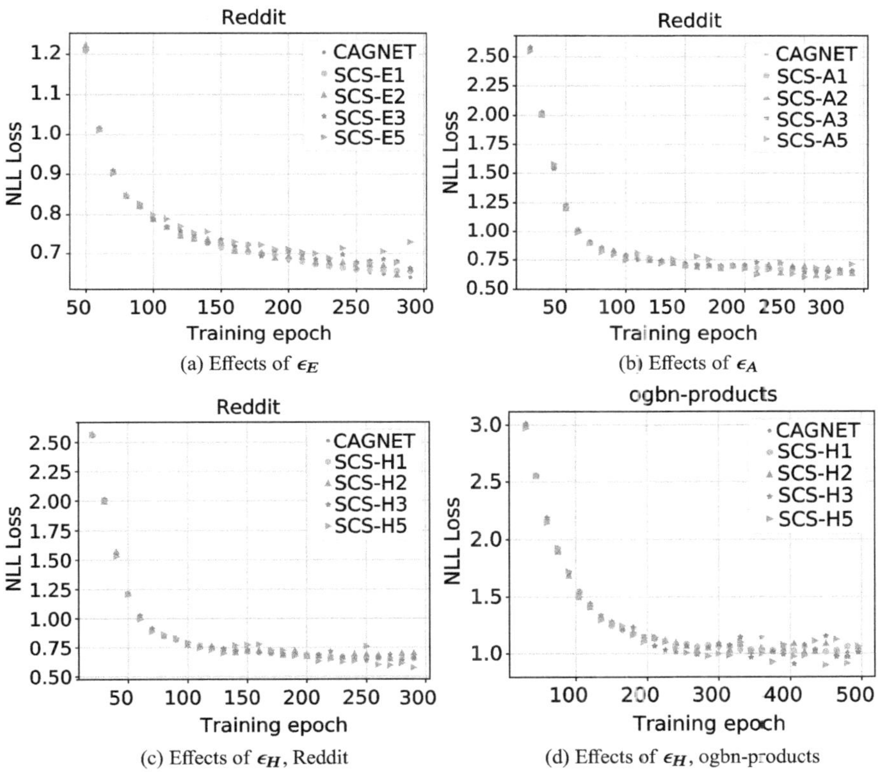

(a) Effects of ϵ_E

(b) Effects of ϵ_A

(c) Effects of ϵ_H, Reddit

(d) Effects of ϵ_H, ogbn-products

Fig. 5.11 Scatter plots of NLL loss with epochs performing broadcast on Reddit and ogbn-products: SCS-E1/E2/E3/E5 denotes $\epsilon_E = \{1, 2, 3, 5\}$; SCS-A1/A2/A3/A5 denotes $\epsilon_A = \{1, 2, 3, 5\}$; SCS-H1/H2/H3/H5 denotes $\epsilon_H = \{0.01, 0.02, 0.03, 0.05\}$

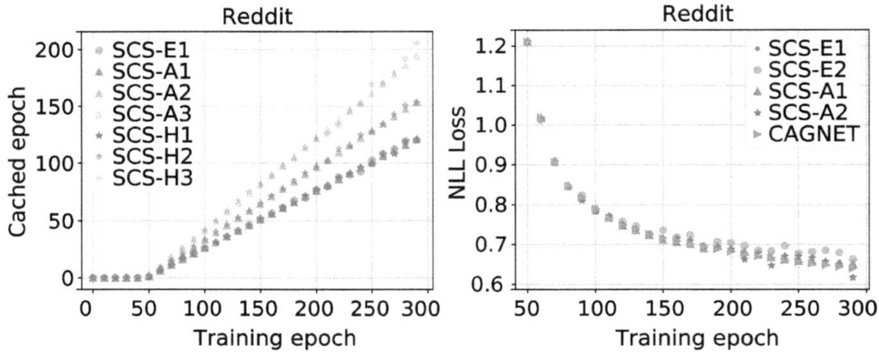

(a) Cached epochs (b) NLL loss

Fig. 5.12 Scatter plots with x-axis denoting epochs performing broadcast on Reddit, to show adaptive staleness-awareness. SCS-E1/E3 denotes $\epsilon_{\mathrm{E}} = \{1, 2\}$; SCS-A1/A2/a3 denotes $\epsilon_{\mathrm{A}} = \{1, 2, 3\}$; SCS-H1/H2/H3 denotes $\epsilon_{\mathrm{H}} = \{0.01, 0.02, 0.03\}$

Table 5.3 Summary of detailed time breakdown (second) measured in seconds in one epoch with SCS-A/E/H compared to CAGNET, on Reddit dataset

Config	Operation	CAGNET	SCS-A1	SCS-E	SCS-H
① 8*1 p2p: no	Compute	0.365	0.359	0.359	0.343
	Communicate	1	0.687	0.697	0.675
② 4*2 p2p: no	Compute	0.093	0.093	0.092	0.090
	Communicate	1	0.717	0.714	0.698
③ 4*1 p2p: yes	Compute	0.437	0.431	0.431	0.425
	Communicate	1	0.703	0.708	0.692

when using all 8 GPUs, its communication cost approaches that of CAGNET running with only 2 GPUs, while also achieving a 67% reduction in computation cost. Notably, the fraction of communication eliminated by Sancus increases with GPU count, a scaling behavior that centralized PS architectures often struggle to match [14]. We also study the effect of model depth in Fig. 5.9c, where Sancus continues to avoid communication across different numbers of GNN layers. Finally, when we enlarge the hidden feature size, Fig. 5.9d indicates that, compared with CAGNET, the communication growth under our epoch-adaptive strategy is both slower and smaller. Together, these results demonstrate the robustness and generality of our framework across diverse system variants.

5.4.3 The Training Efficiency Evaluation

In Table 5.4, we compare end-to-end throughput (epochs/second) of Sancus (SCS-A1) against five representative SOTA distributed GNN systems on their commonly reported Reddit benchmark: CAGNET [4], RoC [3], Dorylus [6], the sampling-based PaGraph [29], and DGCL [2]. RoC introduces a sophisticated memory-management design, and PaGraph performs static caching of high-degree nodes in GPU memory; both rely on non-trivial partitioning to balance workload and reduce cross-device accesses. Dorylus adopts bounded-staleness asynchrony to enable low-cost distributed training with CPU servers, whereas DGCL cuts communication by computing per-node routes optimized for the underlying system topology. For each system, we report the best published configuration that is directly comparable.

Even with our most conservative variant SCS-A1 that yields the smallest communication savings, Sancus still outperforms all compared SOTA systems. SCS-A1 reaches up to 10.3 epochs/second and delivers an average 1.86× throughput improvement. Compared with Dorylus, which targets low-cost full-batch distributed GNN training, Sancus is 68.7× faster while using 80% less cost.

5.4.4 The Model Performance Evaluation

By repeatedly reusing historical embeddings and adaptively invoking Skip-Broadcast, Sancus reduces communication while maintaining GNN accuracy. As shown in Figs. 5.10 and 5.12b, all proposed variants converge to accuracy that is extremely close to CAGNET (within $\leqslant 0.005$), and in some cases matches or even slightly exceeds the baseline despite avoiding communication. Moreover, Sancus typically reaches a satisfactory accuracy level noticeably earlier, indicating faster practical convergence. In contrast, the conventional bounded-gradient approach SkipG exhibits substantially larger accuracy degradation (up to $\leqslant 0.02$) than our embedding-based skip-broadcast strategy. Overall, the experiments demonstrate that Sancus achieves model performance that remains in very close proximity to standard full-graph GNN training, and these empirical observations align with the theoretical results in Sect. 5.3.2.

Sancus Is Staleness-Aware. We next examine how adaptively Sancus skips epochs when reusing cached historical embeddings, and how this adaptivity helps manage system staleness. In Fig. 5.12a, we visualize the epochs at which broadcasting actually occurs during training. As a control, the epoch-fixed variant SCS-E1 broadcasts once every two iterations; consequently, its cached epochs (orange dots) increase in a perfectly regular pattern. In contrast, the epoch-adaptive variants SCS-A and SCS-H exhibit clearly non-uniform broadcast schedules: the gaps between successive dots vary across training, and the irregularity is especially pronounced for SCS-H. This behavior is expected: under the variation-gap criterion, the number of skipped epochs is determined dynamically by the tolerance ϵ_{H}, consistent with the intuition

Table 5.4 Throughput (epochs/second) of GCN on Reddit dataset over different distributed GNN systems

System	Config	Throughput	Reference
SCS-A	V100*4	$1000/97.4 = 10.3$	–
CAGNET	V100*4	$1/0.11 = 9$	Fig. 1 [4]
RoC	P100*4	5	Fig. 5 [3]
Dorylus	Lambda on CPU*2	$130/860 = 0.15$	Sect. 7.2 Table 5.4 [6]
PaGraph	1080ti*4	$1 * 4.9 = 5$	Sect. 5.2 5.5; Fig. 9 [29]
DGCL	V100*4	$\sim (1000/150) = 7$	Fig. 8(a) [2]

Note (1) For RoC result, P100 (4.7 TFLOPs) is 67% as fast as V100 (7 TFLOPS); (2) Dorylus costs 0.085/h * 860 s = $0.2 for 130 epochs, while SCS-A costs 3.06/h * 4 * 13 s = $0.04; (3) PaGraph is sampling-based

in Fig. 5.7. Consistent with Fig. 5.8, the conservative settings SCS-E1/A1/H1 reuse the fewest epochs and therefore yield the smallest communication savings. Crucially, relative to the epoch-fixed SCS-E family, the adaptive strategies achieve higher accuracy (Fig. 5.10), indicating that explicitly managing system staleness better preserves effectiveness. This is further highlighted by SCS-E3: although it caches a comparable number of epochs and achieves similar communication reduction in Fig. 5.8, it incurs the largest accuracy loss in Fig. 5.10. Taken together, these results suggest that staleness-aware adaptivity is both more robust and more advantageous: it enables substantial communication avoidance with little to no accuracy loss, and in some cases even yields slightly better accuracy.

Sancus Is Memory-Efficient. Because the size of GPU memory is a hard resource constraint, the scalability of distributed full-graph GNN training is largely determined by per-GPU memory consumption. In Sancus, GPU memory usage comes from three sources: (i) local data, including embeddings, the local adjacency submatrix, and the replicated weight matrix; (ii) temporary workspace for matrix operations; and (iii) the cache that stores historical embeddings. Relative to CAGNET, the only additional memory introduced by Sancus is this historical-embedding cache. We therefore report the cache footprint in Table 5.5. Overall, the cache accounts for only a small fraction of available memory on modern GPUs for most datasets.

Sancus Is Easy to Generalize. Beyond GCN, we further test generality by applying Sancus to a different full-graph GNN architecture, GAT, in Fig. 5.13a, b. The goal here is to verify that Sancus continues to reduce communication for GAT while maintaining accuracy. We note that the absolute GAT accuracy is not fully optimized, since extensive hyperparameter tuning is outside the scope of this work. We also include results for 1.5D CAGNET, as both 1D and 1.5D are reported to be among the most effective configurations in prior work [4]. While Sancus still cuts communication with minimal accuracy impact, the gains on 1.5D are smaller because the underlying 1.5D algorithm already performs fewer broadcasts. Overall, these results indicate

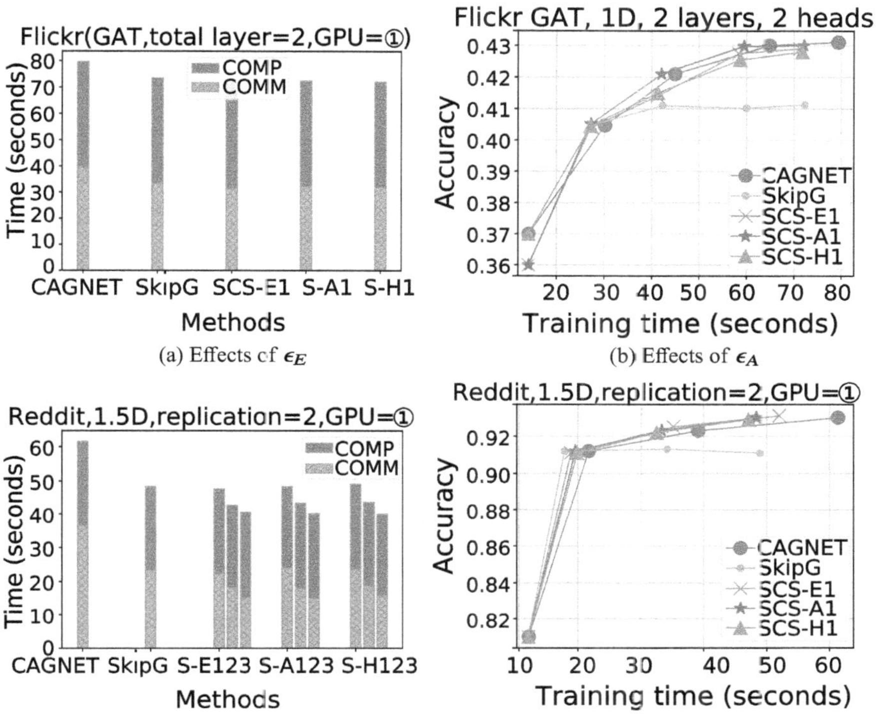

(a) Effects of ϵ_E

(b) Effects of ϵ_A

(c) Effects of ϵ_H

(d) Effects of ϵ_H

Fig. 5.13 Generalization to GAT [30] and 1.5D [4]

that Sancus can be readily extended to other GNN models, and we expect further improvements with more thorough parameter tuning.

5.4.5 *The Effects of the Bounded Staleness*

We further investigate the sensitivity of training to the bounded-staleness hyperparameters in Fig. 5.11, reporting the impact of ϵ_E in Fig. 5.11a, ϵ_A in Fig. 5.11b, and ϵ_H in Fig. 5.11c, d. As the staleness bounds become larger, training trajectories tend to exhibit higher volatility and larger fluctuations. Although a looser bound can slow convergence in terms of iterations, the end-to-end training time can still decrease because more broadcasts are skipped (cf. Fig. 5.10). At the same time, increasing the bounds typically yields greater communication savings (Fig. 5.8) while keeping accuracy loss negligible in our evaluated ranges (Fig. 5.10). Taken together with Figs. 5.8 and 5.10, these results suggest that ϵ_E, ϵ_A, and ϵ_H can be tuned to the needs

Table 5.5 Cache memory footprints with ①

Dataset	2 GPUs	4 GPUs	8 GPUs
Flickr (MB)	170.2	255.3	297.9
Reddit (GB)	0.5	0.8	0.9
Amazon (GB)	1.2	1.8	2.1
ogbn-products (GB)	0.9	1.4	1.6

Ogbn-paper100M is omitted from the table since it only runs on 4 GPUs with 31 GB cache size

of a target application: smaller bounds prioritize accuracy, whereas larger bounds favor higher efficiency.

Understanding the runtime–error tradeoff is inherently challenging in distributed training, because system effects (e.g., communication and stragglers) and learning dynamics (e.g., convergence behavior) are tightly intertwined. In principle, permitting larger staleness can increase throughput, but it may also degrade accuracy [16, 19, 31–33]. In addition, the realized staleness in practice depends on the application, dataset, implementation details, and hardware/software configuration. Figure 5.9 provides a concrete illustration of how such factors can shift behavior. Although datasets react differently to changes in staleness bounds (see Fig. 5.11a, d), across all of our experiments we can still obtain satisfactory model quality with only negligible slowing of convergence. While fully characterizing this tradeoff remains an open research question [31, 32], our results indicate that the adaptive strategies with $\epsilon_A = \{2, 3\}$ and $\epsilon_H = \{0.02, 0.03\}$ provide a practical and effective balance between runtime and accuracy in our GNN training setting.

5.5 Conclusion

In this chapter, we present Sancus, the first staleness-aware, communication-avoiding approach for decentralized GNN training that adaptively reduces communication via historical-embedding caching, while controlling system staleness through bounded embedding staleness and preserving model performance. We introduce three bounded embedding staleness metrics: epoch-fixed staleness, epoch-adaptive staleness, and epoch-adaptive variation-gap staleness. Building on these metrics, Sancus augments decentralized GNN training with a historical-embedding cache and a bounded-staleness checker, enabling GPUs to skip-broadcasts adaptively. In particular, workers reuse cached historical embeddings as long as they remain within the specified bound; once the staleness budget is violated, the latest embeddings are broadcast and the cache is refreshed. We further provide theoretical bounds on both communication cost and approximation error, and conduct extensive experiments on large-scale benchmark graphs to validate the efficiency and effectiveness of Sancus, as well as the importance of adaptivity for managing system staleness. Finally, by reporting memory footprints, we show that caching historical embeddings is memory-efficient.

References

1. Thomas N. Kipf and Max Welling. 2017. Semi-Supervised Classification with Graph Convolutional Networks. In *5th International Conference on Learning Representations, ICLR 2017, Toulon, France, April 24–26, 2017, Conference Track Proceedings*. OpenReview.net. https://openreview.net/forum?id=SJU4ayYgl
2. Zhenkun Cai, Xiao Yan, Yidi Wu, Kaihao Ma, James Cheng, and Fan Yu. 2021. DGCL: an efficient communication library for distributed GNN training. In *EuroSys '21: Sixteenth European Conference on Computer Systems, Online Event, United Kingdom, April 26–28, 2021*. ACM, 130–144.
3. Zhihao Jia, Sina Lin, Mingyu Gao, Matei Zaharia, and Alex Aiken. 2020. Improving the Accuracy, Scalability, and Performance of Graph Neural Networks with Roc. In *Proceedings of Machine Learning and Systems 2020, MLSys 2020, Austin, TX, USA, March 2–4, 2020*. mlsys.org. https://proceedings.mlsys.org/book/300.pdf
4. Alok Tripathy, Katherine A. Yelick, and Aydin Buluç. 2020. Reducing communication in graph neural network training. In *Proceedings of the International Conference for High Performance Computing, Networking, Storage and Analysis, SC 2020, Virtual Event, November 9–19, 2020*. IEEE/ACM.
5. Jianfei Chen, Jun Zhu, and Le Song. 2018b. Stochastic Training of Graph Convolutional Networks with Variance Reduction. In *ICML 2018, Stockholmsmässan, Stockholm, Sweden, July 10–15, 2018(Proceedings of Machine Learning Research)*, Vol. 80. PMLR, 941–949. http://proceedings.mlr.press/v80/chen18p.html
6. John Thorpe, Yifan Qiao, Jonathan Eyolfson, Shen Teng, Guanzhou Hu, Zhihao Jia, Jinliang Wei, Keval Vora, Ravi Netravali, Miryung Kim, and Guoqing Harry Xu. [n.d.]. Dorylus: Affordable, Scalable, and Accurate GNN Training with Distributed CPU Servers and Serverless Threads. In *OSDI 2021, July 14–16, 2021*. USENIX Association, 495–514.
7. Zhao Wang, Yijin Guan, Guangyu Sun, Dimin Niu, Yuhao Wang, Hongzhong Zheng, and Yinhe Han. 2020. Gnn-pim: A processing-in-memory architecture for graph neural networks. In *Conference on Advanced Computer Architecture*. Springer, 73–86.
8. Swapnil Gandhi and Anand Padmanabha Iyer. 2021. P3: Distributed Deep Graph Learning at Scale. In *15th USENIX Symposium on Operating Systems Design and Implementation, OSDI 2021, July 14–16, 2021*. USENIX Association, 551–568. https://www.usenix.org/conference/osdi21/presentation/gandhi
9. Lingxiao Ma, Zhi Yang, Youshan Miao, Jilong Xue, Ming Wu, Lidong Zhou, and Yafei Dai. 2019. NeuGraph: Parallel Deep Neural Network Computation on Large Graphs. In *2019 USENIX Annual Technical Conference, USENIX ATC 2019, Renton, WA, USA, July 10–12, 2019*. USENIX Association.
10. Seungwon Min, Kun Wu, Sitao Huang, Mert Hidayetoglu, Jinjun Xiong, Eiman Ebrahimi, Deming Chen, and Wen-mei W. Hwu. 2021. Large Graph Convolutional Network Training with GPU-Oriented Data Communication Architecture. *Proc. VLDB Endow.* 14, 11 (2021), 2087–2100. https://doi.org/10.14778/3476249.3476264
11. Da Zheng, Chao Ma, Minjie Wang, Jinjing Zhou, Qidong Su, Xiang Song, Quan Gan, Zheng Zhang, and George Karypis. 2020. DistDGL: Distributed Graph Neural Network Training for Billion-Scale Graphs. (2020), 36–44. https://doi.org/10.1109/IA351965.2020.00011
12. Rong Zhu, Kun Zhao, Hongxia Yang, Wei Lin, Chang Zhou, Baole Ai, Yong Li, and Jingren Zhou. 2019. AliGraph: A Comprehensive Graph Neural Network Platform. *Proc. VLDB Endow.* 12, 12 (2019), 2094–2105. http://www.vldb.org/pvldb/vol12/p2094-zhu.pdf
13. Hao Li, Asim Kadav, Erik Kruus, and Cristian Ungureanu. [n.d.]. MALT: distributed data-parallelism for existing ML applications. In *EuroSys 2015, Bordeaux, France, April 21–24, 2015*. ACM.
14. Xiangru Lian, Ce Zhang, Huan Zhang, Cho-Jui Hsieh, Wei Zhang, and Ji Liu. 2017. Can Decentralized Algorithms Outperform Centralized Algorithms? A Case Study for Decentralized Parallel Stochastic Gradient Descent. In *Advances in Neural Information Processing Sys-*

tems 30: Annual Conference on Neural Information Processing Systems 2017, December 4–9, 2017, Long Beach, CA, USA. 5330–5340. https://proceedings.neurips.cc/paper/2017/hash/f75526659f31040afeb61cb7133e4e6d-Abstract.html
15. Amir Gholami, Ariful Azad, Peter H. Jin, Kurt Keutzer, and Aydin Buluç. 2018. Integrated Model, Batch, and Domain Parallelism in Training Neural Networks. In *Proceedings of the 30th on Symposium on Parallelism in Algorithms and Architectures, SPAA 2018, Vienna, Austria, July 16–18, 2018*. ACM, 77–86.
16. James Cipar, Qirong Ho, Jin Kyu Kim, Seunghak Lee, Gregory R. Ganger, Garth Gibson, Kimberly Keeton, and Eric P. Xing. [n.d.]. Solving the Straggler Problem with Bounded Staleness. In *HotOS XIV, Santa Ana Pueblo, New Mexico, USA, May 13–15, 2013*. USENIX Association.
17. George Karypis and Vipin Kumar. 1998. A Fast and High Quality Multilevel Scheme for Partitioning Irregular Graphs. *SIAM J. Sci. Comput.* 20, 1 (1998).
18. Matthias Fey, Jan Eric Lenssen, Frank Weichert, and Jure Leskovec. 2021. GNNAutoScale: Scalable and Expressive Graph Neural Networks via Historical Embeddings. In *ICML 2021, 18–24 July 2021, Virtual Event(Proceedings of Machine Learning Research)*, Vol. 139. PMLR, 3294–3304. http://proceedings.mlr.press/v139/fey21a.html
19. Jiawei Jiang, Bin Cui, Ce Zhang, and Lele Yu. 2017. Heterogeneity-aware Distributed Parameter Servers. In *Proceedings of the 2017 ACM International Conference on Management of Data, SIGMOD Conference 2017, Chicago, IL, USA, May 14–19, 2017*. ACM, 463–478. https://doi.org/10.1145/3035918.3035933
20. Lintao Xian, Bingzhe Li, Jing Liu, Zhongwen Guo, and David H. C. Du. 2021. H-PS: A Heterogeneous-Aware Parameter Server With Distributed Neural Network Training. *IEEE Access* 9 (2021), 44049–44058.
21. Won Kim and Frederick H. Lochovsky (Eds.). 1989. *Object-Oriented Concepts, Databases, and Applications*. ACM Press and Addison-Wesley.
22. Ernie Chan, Marcel Heimlich, Avi Purkayastha, and Robert A. van de Geijn. 2007. Collective communication: theory, practice, and experience. *Concurr. Comput. Pract. Exp.* 19, 13 (2007), 1749–1783.
23. Sergi Abadal, Akshay Jain, Robert Guirado, Jorge López-Alonso, and Eduard Alarcón. 2022. Computing Graph Neural Networks: A Survey from Algorithms to Accelerators. *ACM Comput. Surv.* 54, 9 (2022), 191:1–191:38. https://doi.org/10.1145/3477141
24. Hanqing Zeng, Hongkuan Zhou, Ajitesh Srivastava, Rajgopal Kannan, and Viktor K. Prasanna. 2020. GraphSAINT: Graph Sampling Based Inductive Learning Method. In *8th International Conference on Learning Representations, ICLR 2020, Addis Ababa, Ethiopia, April 26–30, 2020*. OpenReview.net. https://openreview.net/forum?id=BJe8pkHFwS
25. Weihua Hu, Matthias Fey, Marinka Zitnik, Yuxiao Dong, Hongyu Ren, Bowen Liu, Michele Catasta, and Jure Leskovec. [n.d.]. Open Graph Benchmark: Datasets for Machine Learning on Graphs. In *Advances in Neural Information Processing Systems 33: Annual Conference on Neural Information Processing Systems 2020, NeurIPS 2020, December 6–12, 2020, virtual*. https://proceedings.neurips.cc/paper/2020/hash/fb60d411a5c5b72b2e7d3527cfc84fd0-Abstract.html
26. Adam Paszke, Sam Gross, Francisco Massa, Adam Lerer, James Bradbury, Gregory Chanan, Trevor Killeen, Zeming Lin, Natalia Gimelshein, Luca Antiga, Alban Desmaison, Andreas Köpf, Edward Yang, Zachary DeVito, Martin Raison, Alykhan Tejani, Sasank Chilamkurthy, Benoit Steiner, Lu Fang, Junjie Bai, and Soumith Chintala. 2019. PyTorch: An Imperative Style, High-Performance Deep Learning Library. In *NeurIPS 2019, December 8–14, 2019, Vancouver, BC, Canada*. 8024–8035.
27. Xupeng Miao, Xiaonan Nie, Yingxia Shao, Zhi Yang, Jiawei Jiang, Lingxiao Ma, and Bin Cui. 2021. Heterogeneity-Aware Distributed Machine Learning Training via Partial Reduce. In *SIGMOD '21: International Conference on Management of Data, Virtual Event, China, June 20–25, 2021*. ACM, 2262–2270. https://doi.org/10.1145/3448016.3452773
28. Weilin Cong, Rana Forsati, Mahmut T. Kandemir, and Mehrdad Mahdavi. 2020. Minimal Variance Sampling with Provable Guarantees for Fast Training of Graph Neural Networks. In *KDD '20: The 26th ACM SIGKDD Conference on Knowledge Discovery and Data Mining, Virtual*

Event, CA, USA, August 23–27, 2020. ACM, 1393–1403. https://doi.org/10.1145/3394486.3403192

29. Zhiqi Lin, Cheng Li, Youshan Miao, Yunxin Liu, and Yinlong Xu. 2020. PaGraph: Scaling GNN training on large graphs via computation-aware caching. In *SoCC '20: ACM Symposium on Cloud Computing, Virtual Event, USA, October 19–21, 2020*. ACM, 401–415. https://doi.org/10.1145/3419111.3421281
30. Petar Velickovic, Guillem Cucurull, Arantxa Casanova, Adriana Romero, Pietro Liò, and Yoshua Bengio. [n.d.]. Graph Attention Networks. In *6th International Conference on Learning Representations, ICLR 2018, Vancouver, BC, Canada, April 30–May 3, 2018, Conference Track Proceedings*. OpenReview.net. https://openreview.net/forum?id=rJXMpikCZ
31. Wei Dai, Yi Zhou, Nanqing Dong, Hao Zhang, and Eric P. Xing. 2019. Toward Understanding the Impact of Staleness in Distributed Machine Learning. In *ICLR 2019, New Orleans, LA, USA, May 6–9, 2019*. OpenReview.net. https://openreview.net/forum?id=BylQV305YQ
32. Joo Hwan Lee and Hyesoon Kim. 2019. Empirical Investigation of Stale Value Tolerance on Parallel RNN Training. In *ISPASS 2019, Madison, WI, USA, March 24–26, 2019*. IEEE, 153–164. https://doi.org/10.1109/ISPASS.2019.00029
33. Qinyi Luo, Jiaao He, Youwei Zhuo, and Xuehai Qian. 2020. Prague: High-Performance Heterogeneity-Aware Asynchronous Decentralized Training. In *ASPLOS '20: Architectural Support for Programming Languages and Operating Systems, Lausanne, Switzerland, March 16–20, 2020*. ACM, 401–416.

Chapter 6
Scalable Temporal Graph Neural Network Training on Dynamic Graphs

Shihong Gao and Yiming Li

Abstract Temporal Graph Neural Networks (T-GNNs) have become the de facto solution for representation learning on dynamic graphs, enabling state-of-the-art performance on tasks such as temporal link prediction and recommendation. However, existing T-GNN training pipelines suffer from scalability issues due to ill-suited batching and high input data loading costs, which severely limit their efficiency on large-scale graphs. This chapter addresses both these bottlenecks with two complementary system prototypes. First, we propose ETC, a generic framework that introduces a theoretically grounded batch splitting algorithm and a three-step deduplication policy to improve computation throughput and reduce I/O overhead. Second, we present SIMPLE, a dynamic data placement system that maintains a GPU buffer for frequently accessed inputs, optimizing data reuse through an interval selection algorithm with approximation guarantees. Together, ETC and SIMPLE significantly accelerate T-GNN training, achieving up to 62.4× speedup over state-of-the-art baselines while preserving model accuracy, as demonstrated by extensive experiments on real-world datasets.

6.1 Background

Dynamic graphs are fundamental in many settings, such as social networks, citation networks, e-commerce, and recommender systems. They commonly represent real-world processes as time-stamped interactions among nodes. For instance, on social platforms like Reddit, users engage via actions such as commenting or following; in e-commerce, customers interact with items through purchases. Supporting tasks like temporal link prediction and recommendation therefore requires learning representations on evolving graphs, where nodes and edges change over time. In contrast,

S. Gao (✉)
The Hong Kong University of Science and Technology, Hong Kong SAR, China
e-mail: sgaoar@connect.ust.hk

Y. Li
The Hong Kong University of Science and Technology, Hong Kong SAR, China
e-mail: yliix@connect.ust.hk

Y. Shen and L. Chen (eds.), *Graph Neural Network Training*, Machine Learning: Foundations, Methodologies, and Applications,
https://doi.org/10.1007/978-981-95-5795-0_6

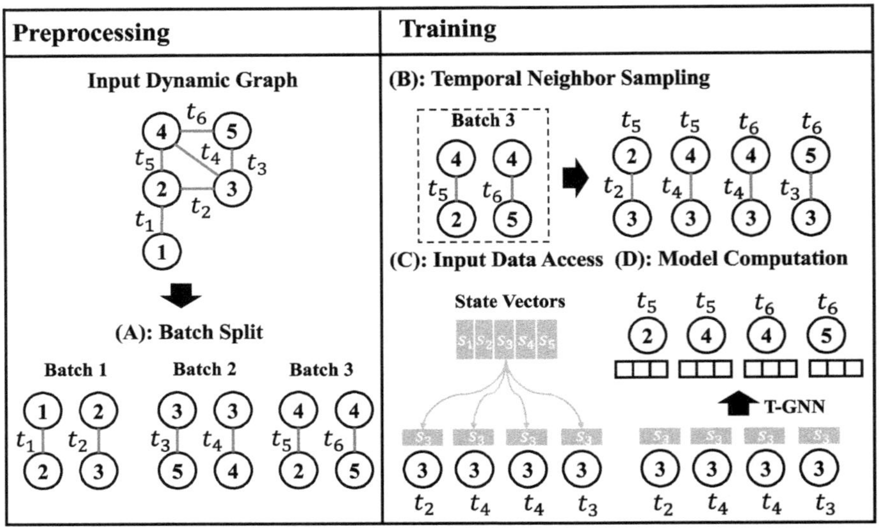

Fig. 6.1 Toy example of T-GNN training. The input dynamic graph has 6 interactions with different timestamps. The batch size is 2, which means that 2 target interactions are contained in each batch. The number of sampled neighbor per node is 1 and the sampling depth is 1

a large body of prior methods [1–4] assumes a static graph. Although these static approaches can be applied to dynamic graphs by ignoring temporal evolution, prior work shows that doing so leads to inferior performance [5, 6].

Temporal Graph Neural Networks (T-GNNs) [5–9] have become a leading paradigm for representation learning on dynamic graphs and have repeatedly delivered state-of-the-art performance. They compute node representations at designated timestamps through recursive temporal message passing. With time-aware neighbor sampling and time-conditioned neighborhood aggregation, T-GNNs can capture how information diffusion evolves over time. Prior work further reports that, on dynamic-graph prediction tasks, T-GNNs substantially outperform both static GNNs [3, 10, 11] and snapshot-based GNN methods [12–14].

In common practice, T-GNNs are trained offline by following the chronological order of interactions [5, 6]. As shown in Fig. 6.1, training starts from an input dynamic graph stored in CPU memory, which is split into multiple batches during preprocessing. Each batch is a timestamp-ascending interaction sequence. To maintain temporal dependencies in the dynamic graph, the GPU processes these batches one after another. For each batch, the system proceeds through three stages: temporal neighbor sampling, input data access, and model computation. Concretely, given the target interactions of a batch, the system first samples temporal neighbors for the participating nodes. It then uses the resulting computation graph to gather the required inputs—e.g., node state vectors summarizing historical interactions—and feeds them into the T-GNN for computation.

Although T-GNNs demonstrate strong performance, their training on large-scale dynamic graphs is extremely resource-intensive–so much so that a single epoch can take several hours to complete [15]. To alleviate this heavy training burden, recent efforts like Orca [16] and Zebra [15] focus on improving the efficiency of the model computation stage. Orca reduces computation by leveraging historical embeddings to skip redundant operations, while Zebra introduces a temporal personalized PageRank-based aggregation scheme that selectively aggregates only the most influential neighbors, thereby lowering the computational cost. However, these optimizations are limited to synchronous T-GNN architectures [5, 6, 9]. In contrast, TGL [17] offers a general-purpose training framework compatible with a wide range of T-GNNs [5, 6, 8, 9, 18], but it primarily accelerates the sampling phase and overlooks other performance-critical stages. We identify two key limitations shared by existing solutions: (1) the small batch size typically used during preprocessing significantly limits model computation efficiency and (2) input data loading consumes a substantial portion of total training time. Together, these bottlenecks hinder the scalability of T-GNN training on large-scale dynamic graphs.

In this chapter, we first propose ETC, a generic framework that addresses key bottlenecks in T-GNN training on large-scale dynamic graphs. It introduces a single-pass batch splitting algorithm that balances efficiency and information retention, with theoretical optimality guarantees. To reduce input data loading costs, ETC further adopts a three-step deduplication policy and a lightweight inter-batch pipeline to streamline data access. Then, to further reduce data loading costs, we present SIMPLE, a T-GNN training system that leverages dynamic data placement. By maintaining a small GPU buffer for frequently accessed inputs, SIMPLE minimizes redundant loading. It formulates a memory-constrained interval selection problem to guide placement and introduces a greedy algorithm with approximation guarantees. Extensive experiments on real-world datasets show that our solutions achieve up to 62.4× training speedup over state-of-the-art methods without sacrificing model performance.

6.2 Efficiency Bottlenecks

In this section, we take a closer look at the two major efficiency bottlenecks identified in Chap. 6.1 that hinder existing T-GNNs from scaling to large-scale dynamic graphs.

6.2.1 *Ill-Suited Batching Scheme in Preprocessing*

The traditional batching strategy used in prior T-GNN training works [6, 8, 15–17] typically adopts small batch sizes, resulting in only a limited number of interactions being processed concurrently on the GPU. This underutilization hampers the computational efficiency of T-GNNs [9]. Simply increasing the batch size within this traditional scheme, however, can amplify the issue of intra-batch information

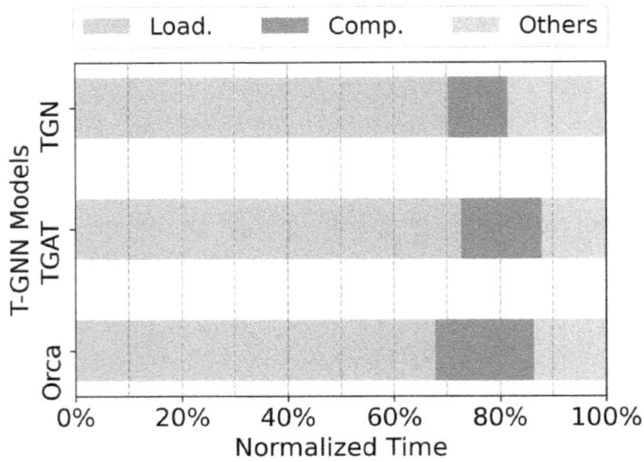

Fig. 6.2 Comparison of the normalized time among the data loading (Load.), model computation (Comp.), and the other data preparation operations including sampling (Others) when training three state-of-the-art T-GNN models on the GDELT dataset [19] under TGL [17] system

loss, as illustrated in Example 1. In practice, while larger batches can improve GPU utilization and speed up training, they often lead to a substantial drop in predictive accuracy. These observations highlight the need for a principled method to quantify intra-batch information loss and a redesigned batching scheme that enables larger batches for improved training efficiency while mitigating information degradation.

Example 1 Consider the two interactions $(v_4, v_2, e_{42}(t_5), t_5)$ and $(v_4, v_5, e_{45}(t_6), t_6)$ in Batch 3 of Fig. 6.1. Since they belong to the same batch and are processed concurrently, they are effectively treated as if they occur simultaneously. Notably, v_4 appears in both interactions. Ideally, T-GNN should update v_4's state vector sequentially at t_5 and t_6. However, because both updates are performed within the same batch, they use the same pre-existing state of v_4. As a result, the update at t_6 fails to incorporate the intermediate state change caused by the interaction at t_5, effectively disregarding it from v_4's temporal perspective.

6.2.2 Overwhelming Data Loading Overhead

As illustrated in Fig. 6.2, input data loading constitutes the primary efficiency bottleneck in T-GNN training on large-scale dynamic graphs, accounting for approximately 70% of the total training time. In conventional pipelines [15–17], input data access is performed by directly retrieving all necessary data based on the sampling results. Specifically, an access list is generated that mirrors the sampled node and edge IDs, and the corresponding data are then transferred from CPU memory to the GPU. While this approach is straightforward, it fails to consider data access patterns in T-GNN training, leading to significant redundancy.

A key observation is that even within one batch, a node may participate in multiple interactions occurring at different timestamps. While the model produces time-specific state vectors and embeddings for that node, these occurrences are still built

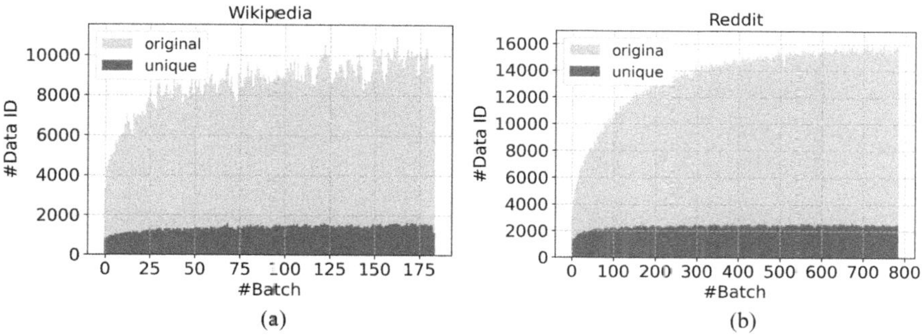

Fig. 6.3 Sampling results on the **a** Wikipedia and **b** Reddit datasets [9]. The X-axis shows the number of batches per epoch, and the Y-axis indicates the number of data IDs. "Original" refers to the total number of sampled data IDs (including duplicates), while "Unique" represents the count of distinct data IDs within the sampling results

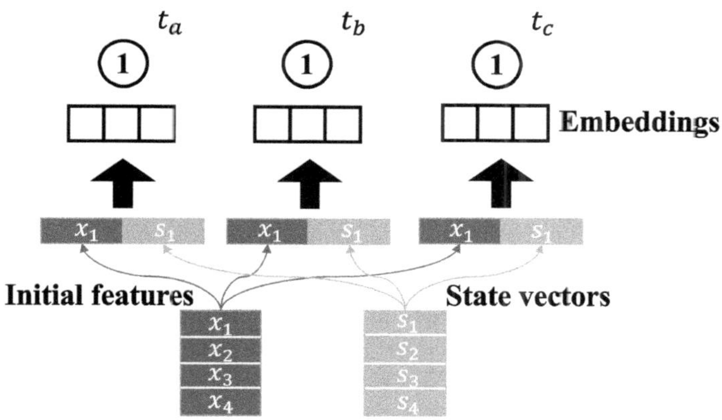

Fig. 6.4 Illustration of data access pattern in T-GNN training

from the same underlying inputs, namely the node features and the node's historical state vectors. Example 2 demonstrates this effect. In real workloads, the resulting redundancy can be pronounced. Figure 6.3[1] shows that repeatedly accessing the same nodes can make redundant data loading account for more than 80% of the overall loading volume.

Example 2 In Fig. 6.4, node v_1 appears at timestamps t_a, t_b, and t_c within the same batch. Although T-GNN generates different embeddings for each timestamp, the same initial feature x_1 and past state vector s_1 are accessed and used in all cases. Under the conventional data access policy, x_1 and s_1 are redundantly selected three times on the CPU and transferred to the GPU. In reality, a single transfer would suffice, as the same input data is reused to generate different embeddings.

[1] Batch size: 600. Sampling strategy: most recent-k neighbor sampling with $k = 10$.

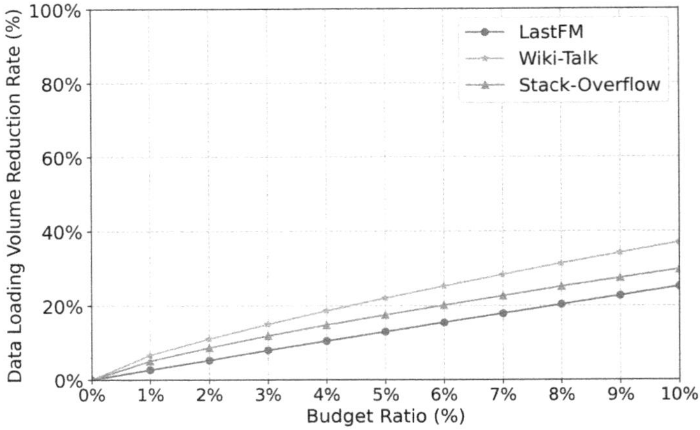

Fig. 6.5 Impact of static data placement strategy for training a state-of-the-art T-GNN model [16] on the LastFM, Wiki-Talk, and Stack-Overflow datasets. The X-axis indicates the buffer budget ratio (GPU buffer size relative to total input data size), while the Y-axis shows the data volume reduction rate (i.e., the proportion of data loading volume eliminated)

6.2.3 Suboptimal Data Placement Strategy

Hybrid CPU-GPU data placement has proven highly effective in accelerating static GNN training [20–22]. By preloading only a small fraction of frequently accessed input data (around 1% to 10% of the total input data footprint) onto the GPU before training, these methods can bypass over 70% of data loading [20]. We aim to adopt this hybrid placement approach for T-GNN training on large-scale dynamic graphs to reduce its substantial data loading overhead. Existing criteria for selecting input data to cache (e.g., nodes with high in-degree [23] or high PageRank scores [24]) focus primarily on static node features. Adapting these to T-GNN, where training is offline, allows precomputing the frequency of each input's occurrence and caching the most frequent ones on the GPU. However, as shown in Fig. 6.5, this static approach yields limited data loading reduction (only 2.7% to 36.9%). This limitation arises because dynamic graphs' temporal aspects often invalidate the assumption of a power-law distribution in occurrence frequency; for instance, in the LASTFM dataset [9], the top 1% most frequent inputs account for merely 4.2% of total occurrences.

In contrast, dynamic hybrid CPU-GPU placement offers promising advantages by incorporating temporal information. By maintaining a GPU buffer whose contents are continually updated based on timestamps, it is possible to achieve greater data loading reduction than static strategies. The challenge lies in managing the GPU buffer effectively, e.g., deciding when and how to update it. Classic strategies such as Least Recently Used (LRU) [25] consider temporal locality by evicting the least recently accessed data. However, ignoring the global occurrence patterns of offline T-GNN training data may limit the effectiveness of purely online methods like LRU.

Indeed, our empirical evaluation of LRU-based buffer updates during T-GNN training shows only marginal improvements over static data placement[2].

6.3 Per-Batch Composition and Data Deduplication

This section presents ETC, a novel T-GNN training framework that accelerates T-GNN training by careful batch split and optimized data loading per graph stream/batch. Specifically, we introduce the details of ETC's information-loss-bounded batch split in Sect. 6.3.1, and ETC's redundancy-aware data loading in Sect. 6.3.2. Finally, we conduct experiments in Sect. 6.3.3.

6.3.1 Information-Loss-Bounded Batching Scheme

Unlike the traditional batching scheme discussed in Sect. 6.2.1, ETC's batching strategy enables larger batch sizes while effectively mitigating information loss. This leads to a substantial improvement in model computation efficiency without sacrificing performance, overcoming the limitations of conventional approaches.

Information-Loss Quantification. To measure the information loss within a batch, we define a novel information-loss score $\beta(B_i)$ for a given batch B_i as:

$$\beta(B_i) = \sum_{v \in N_i} \beta(v, B_i), \tag{6.1}$$

$$\beta(v, B_i) = \left|\{\alpha(t) \in B_i \mid v \in \alpha(t)\}\right| - \mathbf{1}\Big(\left|\{\alpha(t) \in B_i \mid v \in \alpha(t)\}\right| > 0\Big), \tag{6.2}$$

where N_i is the set of nodes involved in batch B_i. Equation 6.2 captures the discrepancy between the actual number of updates and the ideal case (one interaction per node per batch). Equation 6.1 then aggregates the per-node losses to quantify the overall information loss in batch B_i.

From these definitions, we can derive a simplified equivalent form of the batch-level information-loss score:

$$\beta(B_i) = 2|B_i| - |N_i|. \tag{6.3}$$

Problem Formulation. Given a dynamic graph as input, the objective is to divide it into a set of training batches for T-GNNs, such that the overall number of batches is minimized. Since the number of interactions is fixed, this objective is equivalent to maximizing batch sizes. However, we impose a constraint on the information loss

---

[2] Results presented in Chap. 6.4.

of each batch to avoid excessive degradation in model quality. We formalize the problem as follows:

Definition 1 *(Batch Split Problem for T-GNN Training)*Given a set of interactions $\mathcal{G} = \{\alpha(t_1), \alpha(t_2), \ldots, \alpha(t_e)\}$, partition it into K batches $B_1, B_2, \ldots, B_K$, where each batch B_i is a contiguous segment of interactions: $B_i = \cup_{l=p}^{q-1}\{\alpha(t_l)\}$ with $p < q$. Each batch is assigned an information-loss score $\beta(B_i)$. The objective is to find a valid split $f(\mathcal{G}) = \{B_1, B_2, \ldots, B_K\}$ such that:

$$\min_f \quad K$$

$$s.t. \quad \beta(B_i) \leq \varepsilon, \ \forall i, \tag{6.4}$$

$$B_i \cap B_j = \emptyset, \ i \neq j, \ \forall(i, j), \tag{6.5}$$

$$\cup_{i=1}^{K} B_i = \mathcal{G}, \tag{6.6}$$

$$t_m < t_n, \ \forall\alpha(t_m) \in B_i, \ \forall\alpha(t_n) \in B_j, \ i < j \tag{6.7}$$

Constraint 6.4 ensures the information loss in each batch does not exceed a threshold ε. Constraints 6.5 and 6.6 enforce disjointness and completeness of the batch partition. Finally, constraint 6.7 maintains the temporal order of interactions in the original dynamic graph.

Single-pass Batching Algorithm. The batch split problem can be formulated as a time-ordered variant of the classic bin packing problem [26]. To solve it, we develop Algorithm 1, a single-pass procedure that splits batches efficiently. Moreover, we prove theoretically that Algorithm 1 produces an optimal solution for this formulation.

Algorithm 1 first prepares an empty initial batch, an empty set of nodes, and a counter that tracks update occurrences (Lines 1–4). It next traverses the interactions in temporal order (Line 5). For each interaction, the algorithm tentatively inserts it into the active batch and recalculates the resulting information-loss score (Lines 6–10). When the new score still satisfies the information-loss bound, the interaction is kept in the same batch (Lines 11–12); otherwise, the algorithm opens a fresh batch and assigns the interaction to that batch (Lines 13–17). Because the procedure processes each interaction exactly once and decides placement online, the overall runtime is $\mathcal{O}(|E|)$, where $|E|$ is the number of interactions in the dynamic graph.

Theoretical Analysis. This section presents a formal analysis of our single-pass batch splitting algorithm. Despite its simplicity, Algorithm 1 can be shown to attain the optimum for the batch split problem. We begin by introducing two fundamental lemmas that underpin the proof.

Lemma 1 *For a batch $B(p, q)$ that spans from the p-th interaction up to, but not including, the q-th interaction, we have:*

$$\beta(B(p, q)) \leq \beta(B(p, q + c)), \tag{6.8}$$

$$\beta(B(p - c, q)) \leq \beta(B(p, q)), \tag{6.9}$$

Algorithm 1 Single-pass Batch Split Algorithm.

Input: Dynamic graph $\mathcal{G}$; information loss threshold ε.
Output: Batches B_1, B_2, ..., B_K.
Initialize an update counter: $C_u \leftarrow 0$;
Initialize batch ID: $i \leftarrow 0$;
Initialize an empty node set: $N \leftarrow \{\ \}$;
Initialize an empty batch: $B_i \leftarrow \{\ \}$;
for $\alpha(t) \in \mathcal{G}$ **do** ▷ ***Iterate over the interactions sequentially.***
    $C_u \leftarrow C_u + 2$
    for $v \in \alpha(t)$ **do**
        $N \leftarrow N \cup \{v\}$
    end for
    $\beta(B_i) \leftarrow C_u - |N|$
    if $\beta(B_i) \leq \varepsilon$ **then** ▷ ***Stick to the current batch.***
        $B_i \leftarrow B_i \cup \{\alpha(t)\}$
    else ▷ ***State a new batch.***
        $i \leftarrow i + 1$
        $C_u \leftarrow 0$
        $B_i \leftarrow \{\alpha(t)\}$
        $N \leftarrow \{v \in \alpha(t)\}$
    end if
end for

where c is a constant integer.

Proof For the batches $B(p, q)$ and $B(p, q + c)$, denote the node sets in them as $N(p, q)$ and $N(p, q + c)$ respectively. According to the definition of the information-loss score in Eq. 6.3, it follows that:

$$\beta(B(p, q)) = 2(q - p) - |N(p, q)|,$$
$$\beta(B(p, q + c)) = 2(q + c - p) - |N(p, q + c)|.$$

Then we can derive:

$$\beta(B(p, q + c)) - \beta(B(p, q)) = 2c - (|N(p, q + c)| - |N(p, q)|).$$

We know that $N(p, q) \subseteq N(p, q + c)$. Also, $B(p, q + c)$ can be seen as $B(p, q)$ adding c more interactions right after the last interaction it contains. In this way, c more interactions bring at most $2c$ more new nodes to $N(p, q)$. Therefore, we know that:

$$|N(p, q + c)| - |N(p, q)| \leq 2c,$$

which completes the proof of Eq. 6.8. The argument for Eq. 6.9 proceeds analogously.

Lemma 1 demonstrates that the information-loss function is non-decreasing. Concretely, for any batch B, appending more interactions can only make the information-loss score $\beta(B)$ grow, or at best leave it unchanged.

Lemma 2 *Let an optimal solution (OPT) to the batch split problem produce K batches, while Algorithm 1 ($ALG1$) outputs K batches. Let $e(\cdot)$ denote the index of the final interaction contained in a batch. Then, for every $0 < i \leq K$, it holds that $e(i, OPT) \leq e(i, ALG1)$.*

Proof We prove the above lemma via induction. For the base case ($i = 1$), the first batch in both OPT and $ALG1$ begins with the first interaction in the input dynamic graph. By construction, $ALG1$ keeps appending subsequent interactions to this batch and stops only when adding the next one would violate the information-loss constraint. If OPT were to place more interactions into its first batch than $ALG1$ does, that batch would exceed the same constraint, contradicting feasibility. Hence, $e(1, OPT) \leq e(1, ALG1)$.

Assume that for the i-th batch the inequality $e(i, OPT) \leq e(i, ALG1)$ holds. We now show that it also holds for the $(i+1)$-th batch, i.e., $e(i+1, OPT) \leq e(i+1, ALG1)$.

Case1: $e(i+1, OPT) \leq e(i, ALG1)$. Since $ALG1$ produces batches in order, we have $e(i, ALG1) \leq e(i+1, ALG1)$. Combining this with the case condition immediately yields $e(i+1, OPT) \leq e(i+1, ALG1)$.

Case2: $e(i+1, OPT) > e(i, ALG1)$. Because OPT is feasible under the information-loss constraint, its $(i+1)$-th batch does not violate the constraint; in particular, we have:

$$\beta(e(i, OPT) + 1, e(i+1, OPT)) \leq \varepsilon.$$

Then based on Lemma 1, we can obtain that:

$$\beta(e(i, ALG1) + 1, e(i+1, OPT)) \leq \beta(e(i, OPT) + 1, e(i+1, OPT)).$$

From this, it follows that letting the $(i+1)$-th batch of $ALG1$ terminate at $e(i+1, OPT)$ would still satisfy the information-loss constraint. Equivalently, the $(i+1)$-th batch produced by $ALG1$ must extend to at least $e(i+1, OPT)$. Consequently, we obtain $e(i+1, OPT) \leq e(i+1, ALG1)$.

The following Theorem 1 can be proved with Lemma 1 and Lemma 2.

Theorem 1 *Algorithm 1 returns an optimal solution to the batch split problem, i.e., it minimizes the number of batches while satisfying the information-loss score constraint in Definition 1.*

Proof Let $|E|$ be the total number of interactions in the input dynamic graph. Suppose an optimal solution (OPT) uses K^* batches, whereas Algorithm 1 ($ALG1$) outputs K batches, with $K^* \leq K$. Because OPT partitions all interactions into K^* batches, its last batch ends at the final interaction, namely $e(K^*, OPT) = |E|$. By Lemma 2, we have $e(K^*, OPT) \leq e(K^*, ALG1)$. Meanwhile, $ALG1$ scans the interaction sequence once and stops at the last interaction as well, implying $e(K^*, ALG1) = |E|$. Therefore, $ALG1$ cannot use more than K^* batches, and we conclude $K = K^*$.

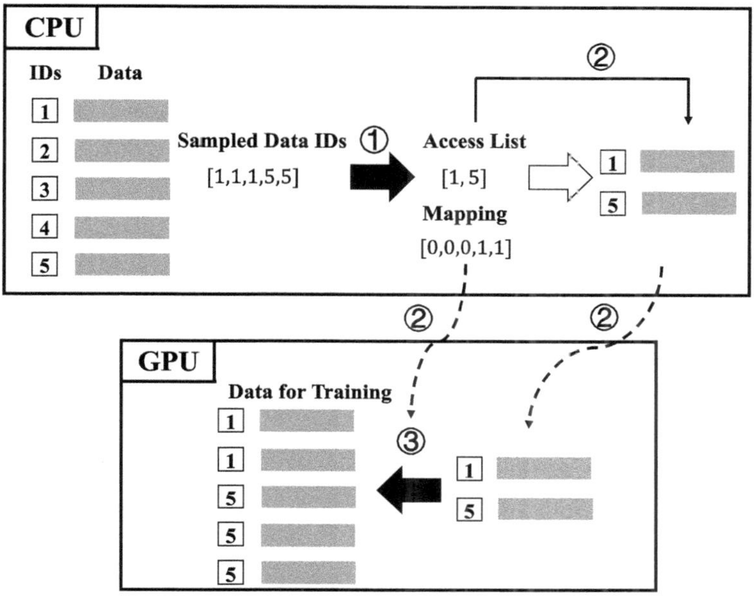

Fig. 6.6 Illustration of the three-step data access policy Supra

6.3.2 *Redundancy-Aware Data Loading*

Inspired by the finding in Sect. 6.2.2, ETC proposes Supra, a new three-stage data access policy, to reduce the overhead of input data access. Figure 6.6 summarizes the workflow. (1) Supra transforms data IDs by deduplicating the sampled IDs into a list of unique IDs, and simultaneously records the mapping from the original IDs to the unique ones. (2) It then uses this unique-ID list to fetch the needed data (node state vectors, node features, and edge features) on the CPU and transfer them to the GPU. (3) After these unique items are available on the GPU, Supra reconstructs the exact input tensors required by the model using the saved mapping.

Data Transfer Volume Analysis. With the conventional access policy, each batch transfers $\mathcal{O}(Q \cdot d)$ data, where (Q) is the number of accessed data IDs and d is the per-item dimensionality (e.g., the node feature size). In contrast, Supra lowers the transfer cost to $\mathcal{O}(R \cdot d + Q)$: it sends only the R unique items (each of dimension d) and additionally transfers the ID mapping, whose size is $\mathcal{O}(Q)$. Because T-GNN training typically has $Q \gg R$, Supra substantially reduces the data transfer volume and thus mitigates transfer overhead.

Inter-Batching Pipeline. To further boost Supra, ETC adds a lightweight pipeline across consecutive batches. Although Supra markedly accelerates input data access, it also incurs extra work in its first stage, where each training batch must be deduplicated to obtain unique data IDs and a mapping back to the original sampled IDs must be constructed. This ID-transformation step runs in (O(Q)) time, with Q being the number of data IDs in a batch. When training on large dynamic graphs with many batches, this per-batch cost accumulates and can erode the benefits of Supra; in extreme cases, the transformation overhead may match or even exceed the savings from transferring less data.

In standard T-GNN training, CPU-side processing and GPU-side computation are arranged in a serial manner: the CPU performs temporal neighbor sampling and prepares inputs, and only then does the GPU execute the model computation. As a result, the CPU is largely idle while the GPU is busy, wasting available resources. We therefore decouple the workloads and overlap them: during GPU training on the current batch, the CPU concurrently performs temporal neighbor sampling and Supra-based data access for the next batch. This overlap hides much of the overhead introduced by Supra. Despite being simple, the inter-batch pipeline further lowers the end-to-end data access cost.

6.3.3 Experiments

Experimental Setups

Implementations. We implement ETC by extending TGL [17] and DGL [27], with PyTorch [28] serving as the underlying training framework for T-GNNs. In particular, we adopt TGL's parallel temporal sampler, which relies on a CSR-style representation to support low-latency temporal neighbor retrieval. For each batch, DGL constructs the Message Flow Graph (MFG) that packages the sampled dynamic subgraph together with the required inputs. In Supra, we use NumPy [29] to efficiently deduplicate data IDs and build the corresponding ID mappings. We realize the inter-batch pipeline via Python threading.

Datasets. Table 6.1 lists four real-world, large-scale dynamic graphs evaluated in our study, each containing millions of interactions. LastFM [9] records one month of listener–music events; Wiki-Talk [30] and Stack-Overflow [31] consist of user-user interactions collected from the corresponding platforms; and GDELT [17] is a near-billion-scale temporal knowledge graph derived from the GDELT 2.0 Event Database [19]. As indicated in Table 6.1, these graphs span a wide range of structural characteristics. For LastFM, Wiki-Talk, and Stack-Overflow, we use the standard chronological splits in [5, 6, 15], with 70%/15%/15% for training/validation/test. For GDELT, we follow [17] and partition interactions occurring before 2019, within 2019, and within 2020 as the training, validation, and test sets, respectively.

Table 6.1 Summary statistics of the dynamic graphs. Here, $|V|$ and $|E|$ are the numbers of nodes and edges, respectively. We use d_v and d_e to indicate the dimensionalities of node and edge features. The symbols α and β refer to the average degree and the graph diameter. We further report Θ, the mean update distance over all nodes. Concretely, for a node v, its average update distance θ_v is defined as the mean number of interactions occurring between any two successive updates of v (averaged across all its updates). The graph-level average update distance is then computed as $\Theta = \sum_{v \in V} \theta_v / |V|$

Dataset	$\|V\|$	$\|E\|$	d_v	d_e	α	β	Θ
LastFM	2 K	1.3 M	128	128	1306	1	2873
Wiki-Talk	1.1 M	7.8 M	172	172	14	11	458149
Stack-Overflow	2.6 M	63.4 M	172	172	49	13	1354646
GDELT	17 K	191.3 M	413	186	22934	7	4876113

Backbone Models. To evaluate ETC, we select three representative T-GNN models:

- **TGAT** [5] uses random Fourier features to encode time and applies an attention mechanism similar to static GNN message passing.
- **TGN** [6] is a general framework encompassing several existing T-GNNs [5, 9, 18], dynamically maintaining node state vectors to capture temporal information.
- **APAN** [8] is an asynchronous, attention-based model that decouples computation and message propagation.

Other T-GNN models [7, 9, 32] can be viewed as extensions of these representatives with more complex designs [33]. Since their training procedures are similar, we exclude their results from the main experiments.

Baseline Frameworks. To validate ETC's efficiency, we compare against three state-of-the-art T-GNN training frameworks:

- **TGL** [17] is a general training framework supporting many T-GNNs, featuring an efficient parallel sampler to tackle temporal neighbor sampling complexity.
- **Orca** [16] addresses model computation bottlenecks via dynamic caching, historical embeddings to reduce computation, and gradient blocking to enhance generalization.
- **Zebra** [15] reduces computation by modifying the aggregation rule to only include the most influential temporal neighbors and accelerates sampling through a single-scan algorithm.

ETC and TGL are model-agnostic frameworks: they preserve the functionality of the underlying T-GNN backbone and instead target data-management optimizations, such as temporal sampling and input access. In contrast, Orca and Zebra enhance efficiency/effectiveness by changing the T-GNN model design; however, they only support synchronous T-GNNs [5, 6, 9] and are implemented for the classic TGN model [6]. Therefore, in the main experiments we report their performance only on the supported TGN backbone. Finally, across all evaluated frameworks (including ETC), the input data are kept in CPU main memory during training.

Training Configurations. We evaluate all models on the widely used link prediction task [5–8, 15, 17], which aims to predict whether future interactions will occur between given node pairs. Since these datasets contain only positive interactions, we generate the same number of negative (i.e., non-existent) links by sampling. Consistent with prior studies [6, 8, 15, 17], we evaluate link prediction using average precision (AP) on the test set.

For neighbor sampling, we employ the top-k recent neighbor strategy [6] with $k = 10$ for all T-GNN models, as it has been shown to yield superior predictive performance [6, 15] compared to other sampling methods.

For TGN and APAN, we use base batch sizes of 1000, 1500, 2000, and 2500 on LastFM, Wiki-Talk, Stack-Overflow, and GDELT, respectively. For fairness, we first run the conventional batching scheme, take the maximum information-loss score among its batches, and use that value as the threshold in our batch split algorithm. In this way, both batching strategies satisfy an identical information-loss constraint. Under this setting, ETC's batch split algorithm yields average batch sizes of 1180, 2180, 2938, and 2635 on the four datasets, respectively.

For TGAT, which lacks the node memory module and thus does not suffer from information loss under large batches, we fix the batch size to 5000 across all datasets and do not apply ETC's batch split algorithm.

We train each model for five epochs on each dataset. To highlight the robustness of our framework, we keep all remaining training hyperparameters identical to the default settings in TGL [17], and do not perform extra tuning. Experiments run on a machine equipped with 96 CPU cores and 256 GB main memory, and training is executed on one NVIDIA RTX 3090 GPU with 24 GB memory. We repeat each experiment three times and report the average outcome.

Main Result

We first conduct an end-to-end comparison of ETC against the baseline frameworks, training identical T-GNN backbones on multiple datasets. Table 6.2 reports the results.

Efficiency of ETC. Compared to existing T-GNN training frameworks, ETC achieves notable training speedups ranging from 1.6× to 62.4×, with most cases surpassing 2×. In particular, against Zebra, ETC delivers 5.4× to 62.4× acceleration. Zebra relies on the conventional data access strategy and suffers from significant overhead due to redundant data transfer, while ETC's Supra policy eliminates much of this redundancy. Compared with Orca, which uses dynamic caching to reduce intermediate computation and indirectly ease data access costs, ETC still yields a 1.6× to 25.7× speedup, owing to its more direct and effective data access optimization. Relative to TGL—the most efficient baseline in terms of sampling—ETC achieves 1.6× to 3.3× faster training. Although TGL's parallel sampler mitigates the cost of temporal neighbor sampling, it neglects the substantial overhead from data access, which ETC explicitly addresses.

Table 6.2 Comparison results for T-GNN training frameworks

Dataset	Model	Framework	AP(%)	Time (s)	Dataset	Model	Framework	AP(%)	Time (s)
LastFM	TGN	Orca	80.11	22.0 (1.6×)	Stack-Overflow	TGN	Orca	**97.64**	12428.7 (25.7×)
		Zebra	**82.44**	76.0 (5.4×)			Zebra	97.57	30167.4 (62.4×)
		TGL	80.92	29.7 (2.1×)			TGL	83.32	1504.4 (3.1×)
		ETC	80.79	**14.1**			ETC	86.10	**483.5**
	TGAT	TGL	67.11	8.5 (2.4×)		TGAT	TGL	87.47	339.4 (1.6×)
		ETC	**67.58**	**3.6**			ETC	**87.55**	**215.9**
	APAN	TGL	**70.05**	19.1 (1.8×)		APAN	TGL	63.10	1696.3 (1.8×)
		ETC	69.79	**10.8**			ETC	**66.42**	**937.4**
Wiki-Talk	TGN	Orca	96.05	343.3 (5.8×)	GDELT	TGN	Orca	TLE	TLE
		Zebra	**96.08**	1029.5 (17.5×)			Zebra	TLE	TLE
		TGL	92.94	135.5 (2.3×)			TGL	98.08	4001.5 (3.3×)
		ETC	93.41	**58.9**			ETC	**98.46**	**1222.4**
	TGAT	TGL	**86.10**	40.7 (2.0×)		TGAT	TGL	**98.08**	1338.2 (2.2×)
		ETC	86.08	**20.1**			ETC	98.07	**619.9**
	APAN	TGL	86.73	160.4 (2.0×)		APAN	TGL	96.62	3302.9 (2.5×)
		ETC	**88.11**	**81.8**			ETC	**96.93**	**1304.5**

Time reports the per-epoch runtime (s). We highlight in bold the highest average precision (%) and the lowest execution time. "TLE" denotes a time-limit exceed, meaning that one training epoch does not complete within 12 h

Effectiveness of ETC. Comparing predictive performance, ETC maintains comparable accuracy to TGL across all T-GNN models, indicating that the efficiency improvements come without compromising model effectiveness. In contrast, Orca and Zebra often achieve higher accuracy for TGN, largely because they modify the TGN model structure to enhance expressiveness and generalization. However, both ETC and TGL adopt the standard TGN architecture without such modifications. Importantly, the optimizations in ETC are orthogonal to those in Orca and Zebra. As a general-purpose framework, ETC can integrate model-side enhancements to further improve performance. In fact, when training the Orca-enhanced TGN model within the ETC framework, we observe significant gains in predictive accuracy. We report these extended results in our technical report [34].

Analysis of ETC

In this section, we perform a direct comparison between our ETC and the baseline framework TGL. As both systems are general-purpose T-GNN training frameworks that accommodate multiple backbone architectures, this study helps quantify the practical gains brought by ETC's optimizations.

Enhanced Model Computation Efficiency. The batch split algorithm in ETC facilitates the use of larger batch sizes when training existing T-GNNs, leading to better GPU utilization and improved computation efficiency. As shown in Fig. 6.7, ETC achieves an average of 1.3× faster model computation time compared to TGL on TGN and APAN, demonstrating the effectiveness of the proposed batching strategy. Moreover, the overhead introduced by the batch split algorithm is minimal—Fig. 6.8 shows it accounts for only about 2.5% of the total end-to-end training time, indicating its practicality.

Accelerated Data Preparation. By combining the three-stage data access policy Supra with a lightweight inter-batch pipeline, ETC substantially lowers the data preparation overhead when training on large-scale dynamic graphs. Figure 6.9 compares the data preparation time of different T-GNN backbones under the two frameworks. Across all settings, ETC is faster than TGL, achieving up to a 5.5× reduction in preparation time and delivering a 3.3× speedup on average.

The Effect of Three-step Data Access Policy Supra. We evaluate the impact of the proposed Supra data access policy using TGN [6] as a representative T-GNN model across all datasets. Results for other T-GNNs show similar trends. As presented in Table 6.3, Supra significantly reduces the data access overhead during data preparation, achieving 4.9×, 4.2×, 3.9×, and 6.1× speedups compared to TGL

Table 6.3 Per-epoch data access time (s) comparison between TGL and ETC

	LastFM	Wiki-Talk	Stack-Overflow	GDELT
TGL	14.2 (4.9×)	66.3 (4.2×)	906.1 (3.9×)	2599.6 (6.1×)
ETC	2.9	15.9	233.3	427.5

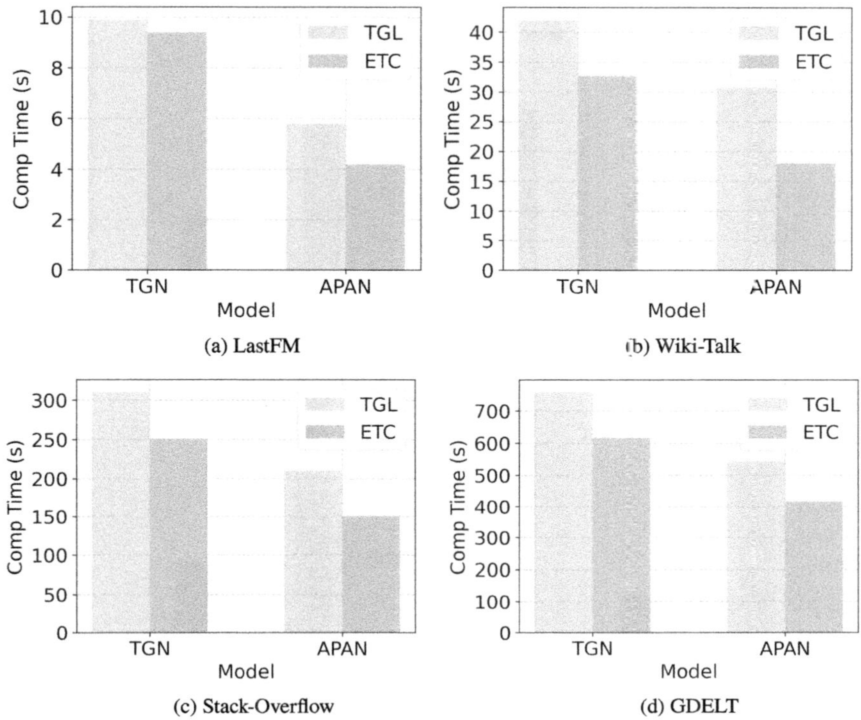

Fig. 6.7 Per-epoch model computation time of TGN and APAN models by ETC and TGL

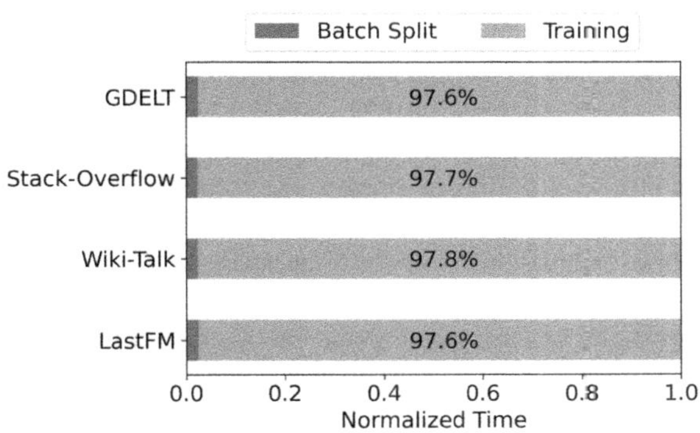

Fig. 6.8 Comparison of normalized time between the batch split preprocessing and the training

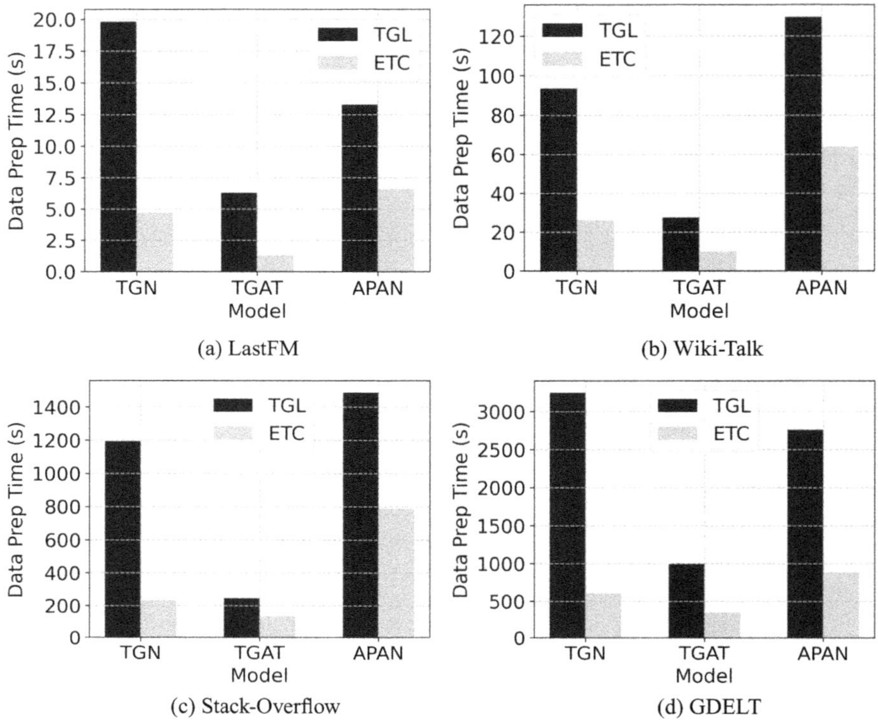

Fig. 6.9 Per-epoch data preparation time comparison between ETC and TGL

across the respective datasets. We also observe that Supra is particularly effective on denser dynamic graphs such as LastFM and GDELT, where the reduction in data access time is more pronounced than on sparser graphs like Wiki-Talk and Stack-Overflow. This is intuitive, as denser graphs tend to introduce more redundant data IDs during neighbor sampling, allowing Supra to eliminate greater redundancy.

Table 6.4 provides a breakdown of the time spent on each of the three steps in Supra. Among them, transferring the unique input data from CPU to GPU is the most time-consuming. The first step, which involves identifying unique data IDs and building reverse mappings, also incurs notable overhead. In contrast, the third step, where the original input data are reconstructed on GPU using the mapping, contributes negligible additional cost.

The Impact of the Inter-batch Pipeline Mechanism. We evaluate the impact of the lightweight inter-batch pipeline in ETC. Using TGN as the backbone, we run experiments on all datasets and compare the training efficiency of ETC with the pipeline enabled versus disabled. As reported in Table 6.5, the pipeline delivers an extra 1.2×–1.6× speedup in training. These results indicate that overlapping CPU-side preparation with GPU computation further reduces overhead and improves the end-to-end efficiency of ETC.

Table 6.4 Per-epoch time (s) comparison for three steps of the data access policy Supra

	LastFM	Wiki-Talk	Stack-Overflow	GDELT
Step I	2.1	10.3	137.3	276.2
Step II	2.9	15.9	233.3	427.5
Step III	0.7	2.5	22.2	60.5

Table 6.5 Per-epoch time (s) comparison of ETC with and without the inter-batch pipeline

	LastFM	Wiki-Talk	Stack-Overflow	GDELT
w/o pipeline	16.9 (1.2×)	75.7 (1.3×)	783.6 (1.6×)	1850.0 (1.5×)
w pipeline	14.1	58.9	483.5	1222.4

6.4 Cross-Batch Dynamic Data Placement

6.4.1 Overview

In this section, we present SIMPLE, the first general-purpose T-GNN training system equipped with a GPU buffer. Designed for efficient training over large-scale dynamic graphs, SIMPLE employs a dynamic hybrid data placement strategy across CPU and GPU. It stores memory-intensive input data–such as edge features–in CPU main memory, while maintaining a flexible GPU buffer to accommodate dynamically changing hot data. We further detail the required preprocessing steps in Sect. 6.4.2 together with by an explanation of the complete training workflow and key components in Sect. 6.4.3.

6.4.2 Preprocessing

Preprocessing. SIMPLE performs preprocessing on the CPU using a sampler and a plan generator. It begins by analyzing data access patterns expected during training through pre-sampling on the input dynamic graphs. For each data batch, the sampler executes temporal neighbor sampling for all root nodes involved in the target interactions (edges), based on their associated timestamps. To enable efficient sampling, SIMPLE integrates a parallel C++-based sampler from [17]. By default, it adopts the widely used top-*k* recent neighbor sampling strategy [6, 16, 17], as described in Chap. 2.3. Prior studies [6, 8, 16, 17] have shown this strategy to outperform other alternatives [5], as it leverages temporal information during sampling, leading to better predictive performance in T-GNN models.

The sampler outputs the IDs of required input data across all batches, which are then passed to the plan generator. The plan generator extracts data access patterns from these IDs and produces GPU buffer plans for each batch. Each buffer plan

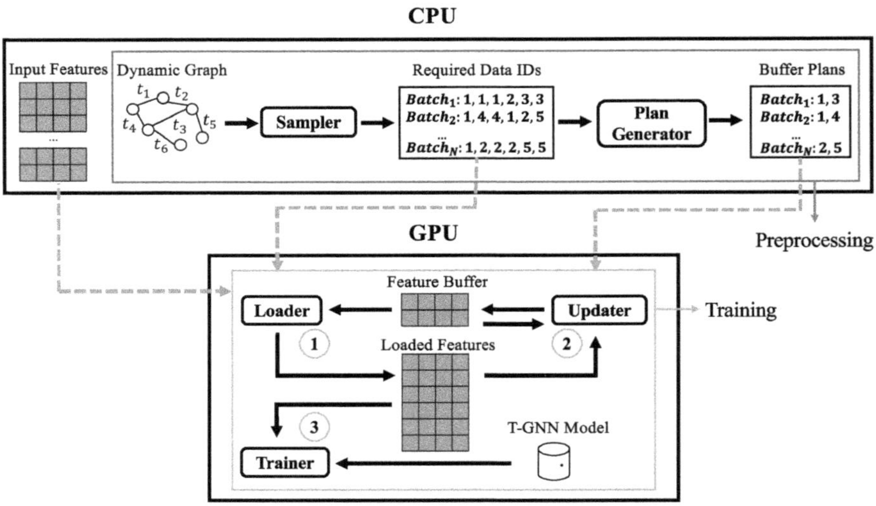

Fig. 6.10 Overall architecture of SIMPLE

Algorithm 2 T-GNN Training by SIMPLE for a Single Epoch.

Input: The number of training batches N; the total CPU-side input data H; the required data ID list for all batches D=[$d_{(1)}$, $d_{(2)}$, $\cdots$, $d_{(N)}$]; the buffer plans for all batches P=[$p_{(1)}$, $p_{(2)}$, $\cdots$, $p_{(N)}$]; the GPU buffer $\xi_{(0)}$; the input untrained T-GNN model $\theta_{(0)}$.

Output: Trained T-GNN model $\theta_{(N)}$.

1: **for** $i \leftarrow 1 : N$ **do**
2: $\quad H_{(i)} \leftarrow \mathsf{LOAD}(d_{(i)}, \xi_{(i)}, H)$ ▷ ***CPU-GPU data loading.***
3: $\quad \xi_{(i+1)} \leftarrow \mathsf{UPDATE}(p_{(i)}, \xi_{(i)}, H_{(i)})$ ▷ ***GPU buffer update.***
4: $\quad \theta_{(i)} \leftarrow \mathsf{TRAIN}(\theta_{(i-1)}, H_{(i)})$ ▷ ***T-GNN model training.***
5: **end for**

---

specifies the set of input data IDs that should be kept in GPU memory for that batch. Further details on the plan generator's functionality are provided in Sect. 6.4.4.

6.4.3 Training Workflow

The system architecture of SIMPLE is illustrated in Fig. 6.10. Since SIMPLE adopts dynamic data placement during T-GNN training, the standard training workflow must be adapted accordingly. Algorithm 2 outlines the modified training procedure for a single epoch under SIMPLE. Each batch involves three key steps executed sequentially: batched data loading (Line 2), buffer update (Line 3), and model computation (Line 4). Importantly, temporal neighbor sampling is not performed during training, as the required data IDs for all batches have already been precomputed during preprocessing. These three steps are handled by SIMPLE's GPU-side modules: the loader, updater, and trainer, described below.

Loader. For each batch $B_{(i)}$, given the required input data IDs $d_{(i)}$, the loader retrieves the corresponding data $H_{(i)}$ from either CPU main memory (H) or the GPU buffer $(\xi_{(\cdot)})$. Its implementation follows the design of loaders in static GNN systems [20–22], with additional GPU-side buffer space allocated for metadata. This metadata—a hash table—enables quick lookups to check whether specific data reside in the GPU buffer and, if so, their exact locations. Leveraging this metadata, the loader achieves efficient ID-based access, with lookup overhead typically accounting for only 1âŁ"-2% of total training time.

The loader's time complexity is $\mathcal{O}(U_{(i)})$, where $U_{(i)}$ denotes the number of data points that must be fetched from CPU memory for batch $B_{(i)}$. Thanks to the GPU buffer, a significant portion of data can be directly accessed on the GPU, reducing the loading cost substantially (i.e., $U_{(i)} \ll |d_{(i)}|$). We will show later in the experiment section that SIMPLE's buffer management scheme can reduce data loading overhead by more than 80%.

Updater. Once the required input data $H_{(i)}$ for batch $B_{(i)}$ have been loaded onto the GPU, the updater is responsible for maintaining and updating both the main contents (i.e., stored input data) and the metadata of the GPU buffer. This update is guided by three inputs: the buffer plan $p_{(i)}$, the freshly loaded data $H_{(i)}$, and the currently buffered data $\xi_{(i)}$. Since all buffer plans are precomputed during the preprocessing stage, the updater can directly retrieve the plan for batch $B_{(i)}$ from CPU memory. Using this plan, the updater efficiently tracks and swaps data between the newly loaded inputs and the existing GPU buffer content.

The updated GPU buffer is then used during data loading for the subsequent batch. Unlike runtime-based policies such as LRU [25], pre-generating buffer plans is advantageous in T-GNN training scenarios, which often involve repeated fine-tuning. These static plans can be reused across training runs, reducing repeated overhead. To further optimize performance, the entire buffer update process is executed on the GPU to avoid unnecessary data transfer costs.

The complexity of the buffer update step for batch $B_{(i)}$ is $\mathcal{O}(Z_{(i)})$, where $Z_{(i)} = |d_{(i)}| + |p_{(i)}|$ represents the total number of data points in GPU memory after loading. Our optimized GPU-side implementation ensures that the buffer update incurs less than 10% of the total training time on average across all benchmarks, with detailed analysis provided later in the experiments.

Trainer. Once the buffer update completes, the trainer module proceeds with model computation using the already loaded data $H_{(i)}$. The core computational process of T-GNNs has been described in Chap. 2.3. SIMPLE includes ready-to-use implementations of state-of-the-art T-GNN models such as [5, 6, 16]. For users who wish to train their own T-GNN models on large-scale dynamic graphs, they only need to provide their custom model logic to the trainer. The remaining components including preprocessing, data loading, and buffer management are automatically handled by SIMPLE.

A typical T-GNN performs recursive temporal aggregation, with time complexity for batch $B_{(i)}$ given by $\mathcal{O}(V_{(i)} \cdot \eta^L)$, where $V_{(i)}$ is the number of target nodes (with

Data Occurrence Patterns

Data IDs	$Batch_1$	$Batch_2$	$Batch_3$	$Batch_4$	$Batch_5$
❶	5	5	0	0	2
❷	1	0	6	2	0
❸	2	4	0	4	2
❹	0	1	6	1	0

Interval A

$Batch_1$ 8 $Batch_4$

❷ → ❷

Schedule:
Retain the vector data 2 (e.g., edge feature vector) in GPU buffer from $Batch_1$ to $Batch_4$.

Profit:
Avoid 8 times (6 in $Batch_3$ + 2 in $Batch_4$) of loading for vector data 2 across $Batch_1$ to $Batch_4$.

Fig. 6.11 Exemplar visualization of the collected data occurrence patterns, where rows and columns denote the data IDs and batch IDs respectively, and each entry further denotes the number of occurrences of a single data point in an individual batch. According to such collected patterns, we can measure the profit of buffering a certain data point over a batch interval (as exemplified in interval A)

timestamps) in the batch, η is the sampling fanout, and L denotes the number of message-passing layers.

Summary. In contrast to standard T-GNN training workflows [17], SIMPLE introduces two key preprocessing enhancements: (1) materializing neighbor sampling results and (2) generating GPU buffer plans in advance. During training, SIMPLE further distinguishes itself by leveraging an efficient GPU-buffer-based data loading mechanism along with a lightweight buffer update step.

6.4.4 Dynamic Data Placement

In this section, we detail how SIMPLE manages the GPU buffer to support efficient T-GNN training. The core challenge lies in designing effective dynamic data placement strategies referred to as buffer plans that minimize data loading overhead. To address this, we first extract global data access patterns through pre-sampling. Using these patterns, we formulate the problem as an interval selection problem and develop a greedy algorithm with approximation guarantees to solve it. The selected intervals represent the desired data placement strategy across training batches, based on which we efficiently generate executable buffer plans for the GPU.

Global Occurrence Patterns. During preprocessing, the sampler in SIMPLE performs a one-time pre-sampling step to capture global data access patterns across all training batches. These patterns are organized into a tabular format, as illustrated in Fig. 6.11. Notably, the captured patterns reflect not only which data points appear in each batch, but also how many times each one occurs. This distinction is important due to a redundancy issue previously observed in data loading [17], where the same data point may be repeatedly loaded even within a single batch. This typically happens when the T-GNN model requires the same node identified by the same ID at different timestamps to generate embeddings. By incorporating this redundancy information, the global data occurrence patterns enable more effective dynamic data placement, ultimately reducing data loading overhead.

Interval Selection Problem. We formalize an optimization problem for designing dynamic data placement strategies tailored to T-GNN training. At a high level, a dynamic data placement strategy defines a temporal retention schedule for each input data point. Specifically, it determines a batch interval during which a particular data point remains resident in the GPU buffer.

Leveraging the global data occurrence patterns collected in the preprocessing stage, we can evaluate the benefit or profit of keeping a data point in GPU memory over a specific batch interval. Each such interval that offers measurable savings in data loading is referred to as a *candidate batch interval*, formally defined as follows:

Definition 2 *(Candidate Batch Interval)* Given a batch interval $\sigma(\cdot)$ with start batch $s_{(\cdot)}$, end batch $e_{(\cdot)}$, and a data point x, we define $\sigma(\cdot)$ as a *candidate batch interval* if retaining x in the GPU buffer from $s_{(\cdot)}$ to $e_{(\cdot)}$ results in a reduction of $p_{(\cdot)}$ in data loading volume for x at $e_{(\cdot)}$. This requires that x is already present in GPU memory at the start of the interval, i.e., at $s_{(\cdot)}$.

The value $p_{(\cdot)}$ is referred to as the *profit* of buffering x over interval $\sigma(\cdot)$. The *size* $h_{(\cdot)}$ denotes the memory footprint of x, and the *makespan* $\Delta(\cdot)$ of the interval is defined as the length of time x is held in GPU memory, calculated as $\Delta(\sigma(\cdot)) = e_{(\cdot)} - s_{(\cdot)}$.

A candidate batch interval is thus represented as a tuple:

$$\sigma(\cdot) = (s_{(\cdot)}, e_{(\cdot)}, x, p_{(\cdot)}, h_{(\cdot)}),$$

indicating the start and end batches, the data ID, the profit of retaining x, and its memory size.

We now describe how to extract all candidate batch intervals from the global data occurrence patterns. Take the example shown in Fig. 6.11: consider the target data point with ID 2 (with size denoted as $h_{(2)}$). According to the occurrence pattern, this data point appears 1 time in batch 1, 6 times in batch 3, and 2 times in batch 5. Following Definition 1, we can derive two candidate batch intervals for this data point:

$$(1,\ 3,\ 2,\ 6 \cdot h_{(2)},\ h_{(2)}) \quad \text{and} \quad (3,\ 5,\ 2,\ 2 \cdot h_{(2)},\ h_{(2)}),$$

where each interval denotes a period during which retaining the data point in GPU memory yields a profit proportional to its number of accesses multiplied by its size. Importantly, we only consider intervals that cannot be decomposed into combinations of existing candidates as valid. As a result, if a data point appears in m distinct batches, we generate exactly $m - 1$ candidate intervals. In other words, for each data point, the candidate intervals are non-overlapping in terms of their active batch ranges.

After extracting all candidate batch intervals from the global data occurrence patterns, we now formulate an optimization problem that aims to select a subset of these intervals to maximize the total reducible data loading volume. This problem is subject to the constraint that, within every timespan between consecutive batches, the total size of data buffered on the GPU cannot exceed a fixed capacity. We define this optimization task as the Interval Selection Problem for T-GNN Training, formally presented in Definition 3.

Definition 3 *(Interval Selection Problem for T-GNN Training (ISPT))* Given n candidate batch intervals $\{\sigma(1), \sigma(2), \ldots, \sigma(n)\}$, where each interval $\sigma(i)$ is associated with a size s_i and a profit p_i, the goal is to select a subset of intervals that maximizes the total profit while satisfying GPU buffer constraints.

Let T denote the total number of timespans (i.e., batch gaps), and let k be the available GPU buffer size. The optimization problem is formulated as follows:

$$\max. \quad \sum_{i=1}^{n} p_i x_i,$$

$$\text{s.t.} \quad \sum_{i=1}^{n} \overline{s}_{it} x_i \leq k, \quad \forall t = 1, 2, \ldots, T, \tag{6.10}$$

$$\overline{s}_{it} \in \{0, s_i\}, \quad \forall i, t, \tag{6.11}$$

$$x_i \in \{0, 1\}, \quad \forall i = 1, 2, \ldots, n, \tag{6.12}$$

where $\overline{s}_{it}$ indicates whether the i-th interval occupies the t-th timespan: $\overline{s}_{it} = s_i$ if $\sigma(i)$ spans the timespan t, and $\overline{s}_{it} = 0$ otherwise.

Hardness. It is straightforward to observe that the classic 0-1 Knapsack problem [35–37] is a special case of ISPT when $T = 1$, i.e., when the training process involves only two batches. This directly implies that the ISPT problem is NP-hard.

Two-Step Interval Selection. To tackle the ISPT problem defined in Definition 3, SIMPLE adopts an efficient two-phase strategy. In the first phase, it distributes the buffer budget across different categories of intervals. Then, in the second phase, it independently solves the ISPT problem within each category using a greedy algorithm.

The plan generator in SIMPLE begins by allocating separate buffer budgets for two categories of input data required in T-GNN training: node-related data (including updatable node state vectors and static node features) and edge-related data (i.e., static edge features). Given a total GPU buffer size budget k, the portion allocated to node-related intervals is computed as:

$$k_V = k \times \frac{vol(V)}{vol(V) + vol(E)}, \tag{6.13}$$

where $vol(V)$ and $vol(E)$ represent the total profit (i.e., the cumulative savable loading volume) of node-related and edge-related intervals, respectively. This allocation strategy reflects the intuition that data types contributing more to potential load-

Algorithm 3 One-pass Interval Selection by SIMPLE

Input: Interval list $L = [\sigma(1), \sigma(2), \ldots, \sigma(n)]$; temporal net profit list $F = [f(\sigma(1)), f(\sigma(2)), \ldots, f(\sigma(n))]$; number budget $\overline{k}_{(\cdot)}$; number of timespans T.
Output: Selected interval list $\tilde{L}$.

Initialize an empty list $\tilde{L} \leftarrow [\,]$
Initialize budget vector $b \leftarrow [\overline{k}_{(\cdot)}, \ldots, \overline{k}_{(\cdot)}]$, where $b \in \mathbb{N}^T$
Sort L in descending order of F, obtaining sorted list $\hat{L}$
for $\sigma(\cdot)$ in $\hat{L}$ **do**
  $s \leftarrow \sigma(\cdot)[0]$, $e \leftarrow \sigma(\cdot)[1]$
  Extract subvector $r \leftarrow b[s : e + 1]$
  for $j \in \{0, 1, \ldots, e - s\}$ **do**
    $r[j] \leftarrow r[j] - 1$
  end for
  if $r[j] \geq 0$ for all j **then**
    $b[s : e + 1] \leftarrow r$
    Append $\sigma(\cdot)$ to $\tilde{L}$
  end if
end for

ing overhead should receive proportionally more buffer budget to maximize overall savings.

Similarly, the budget for edge-related intervals is computed using the same proportional scheme. After determining the size budget $k_{(\cdot)}$ for each data type, we further convert it into a numerical constraint on the maximum number of overlapping intervals allowed per timespan. This is expressed as:

$$\overline{k}_{(\cdot)} = \frac{k_{(\cdot)}}{h(\cdot)}, \tag{6.14}$$

where $h(\cdot)$ denotes the memory size (dimension) of the corresponding input data. As a result, the original size constraint is reformulated as a count-based constraint, indicating that no more than $\overline{k}_{(\cdot)}$ intervals of the same type may overlap within any given timespan.

After budget allocation, SIMPLE's plan generator performs interval selection for each category of input data in parallel, adopting a greedy strategy. To guide this selection process, we introduce a novel metric called the temporal net profit $f(\cdot)$, which quantifies the efficiency of each interval $\sigma(\cdot)$ in terms of profit gained per unit time. It is formally defined as:

$$f(\sigma(\cdot)) = \frac{p(\sigma(\cdot))}{\Delta(\sigma(\cdot))}, \tag{6.15}$$

where $p(\sigma(\cdot))$ and $\Delta(\sigma(\cdot))$ represent the profit and makespan of interval $\sigma(\cdot)$.

Using this metric, the plan generator conducts a one-pass greedy interval selection procedure for each data type, as outlined in Algorithm 1.

Algorithm 3 begins by initializing an empty list of selected intervals $\tilde{L}$ and a budget vector b with length T, the number of timespans between batches (Lines 1–2). Each element of b is set to the per-timespan number constraint $\overline{k}_{(\cdot)}$ as derived in Eq. 6.14. The algorithm then sorts the candidate intervals in descending order based on their temporal net profit values (Line 3).

Next, it iteratively evaluates each interval $\sigma(\cdot)$ in the sorted list $\hat{L}$ (Line 4). For a given interval, the algorithm retrieves its start and end batch indices (Line 5) and extracts the corresponding subvector from the budget vector (Line 6). If selecting the interval does not violate the budget constraint in any covered timespan (Line 10), the budget vector is updated accordingly (Line 11), and the interval is added to the output list $\tilde{L}$ (Line 12). The constraint ensures that GPU memory usage never exceeds the per-timespan budget, thereby avoiding memory overflow.

Intuition. The rationale behind this greedy strategy is to prioritize intervals that offer high profit relative to their makespan. By selecting such intervals first, SIMPLE maximizes memory efficiency, allowing more profitable intervals to fit within the limited GPU buffer over time, thereby minimizing overall data loading cost.

Time Complexity. The sorting step (Line 3) requires $\mathcal{O}(n \log n)$ time. The main loop iterates over all n intervals, and for each, performs a bounded check over its timespan (with average length $\overline{m} < T$ in practice). Thus, the overall complexity of the interval selection procedure is $\mathcal{O}(n \cdot \overline{m})$.

Approximation Guarantee. We also develop the approximation guarantee of SIMPLE on the achieved profits of dynamic data placement (ISPT problem described in Definition 3, i.e., the achieved reduction of data loading volume in T-GNN training by SIMPLE as shown in Line 2 of Algorithm 2). Such an approximation guarantee is summarized in Theorem 2.

Lemma 3 *For the uniform size ISPT problem (single type-d intervals), denote the number budget (how many intervals can overlap in every timespan) and its upper bound as $\overline{K}_d$ and $\overline{K}^*_d$. Denote the result by any feasible solution under $\overline{K}_d$ as $\Phi^d_x(\overline{K}_d)$, we have:*

$$\Phi^d_{OPT}(\overline{K}_d) \geq vol(d)\frac{\overline{K}_d}{2\overline{K}^*_d}, \tag{6.16}$$

where $vol(d) = \Phi^d_{OPT}(\overline{K}^*_d)$. It represents the maximal sum of profits of all type-d intervals (the maximal profits that can be gained via dynamic data placement for type-d input data).

Proof We prove Lemma 1 by contradiction. Given two number budgets $\overline{K}_d(i)$ and $\overline{K}_d(j)$ ($\overline{K}_d(i) \leq \overline{K}_d(j)$, and $\overline{K}_d(i), \overline{K}_d(j) \in N^+$), if $\frac{\overline{K}_d(j)}{\overline{K}_d(i)} = c$ and $c \in N^+$, assume $\frac{\Phi^d_{OPT}(\overline{K}_d(j))}{\overline{K}_d(j)} > \frac{\Phi^d_{OPT}(\overline{K}_d(i))}{\overline{K}_d(i)}$, then we have:

$$\frac{\Phi_{OPT}(\overline{K}_d(j))}{c} > \Phi_{OPT}(\overline{K}_d(i)). \tag{6.17}$$

Note that $\Phi_{OPT}(\overline{K}_d(j))$ can be decomposed into several feasible solutions for ISPT problem with budget smaller than $\overline{K}_d(j)$, we can get:

$$\Phi^d_{OPT}(\overline{K}_d(j)) = c\Phi^d_x(\overline{K}_d(i)), \tag{6.18}$$

where x denotes any feasible solution. Based on Eq. 6.17 and the pigeonhole principle, we can know that there exists at least one $\Phi^d_x(\overline{K}_d(i))$ that $\Phi^a_x(\overline{K}_d(i)) > \Phi^d_{OPT}(\overline{K}_d(i))$, which contradicts to the fact that $\Phi^d_{OPT}(\overline{K}_d(i)) = max(\Phi^d_x(\overline{K}_d(i)))$. Thus the initial assumption is false when $c \in N^+$. Let $K_d(j) = K^*_d$, we can obtain Eq. 6.16 through simple rounding operations under the general case ($c \in R^+$).

Theorem 2 *On ISPT problem with two types of intervals (the general case under the T-GNN training setting), denote the size budget as K, and the result by any feasible solution as $\Phi_x(K)$. For the optimal solution and the solution by SIMPLE denoted as OPT and ALG, we have:*

$$\frac{\Phi_{ALG}(K)}{\Phi_{OPT}(K)} \geq \frac{K}{4MT}, \tag{6.19}$$

where M is the size sum of all intervals (total memory footprint of the input data in T-GNN training), and T represents the total timespan.

Proof Denote the node-related interval size, and the number budget for node-related intervals by SIMPLE as h_V and $\overline{K}_V$. The same definitions for edge-related intervals are represented by h_E and $\overline{K}_E$. Denote M as the total memory footprint, so that $M = \overline{K}^*_V h_V + \overline{K}^*_E h_E$. Then combine Lemma 1, Theorem 1, the budget allocation principle defined in Eq. 6.13, and the fact that $\Phi_{OPT}(K) \leq vol(N) + vol(E)$, we have:

$$\frac{\Phi_{ALG}(K)}{\Phi_{OPT}(K)} \geq \frac{vol(N)\overline{K}_N}{2T\overline{K}^*_V} + \frac{vol(E)\overline{K}_E}{2T\overline{K}^*_E} \tag{6.20}$$

$$\geq \frac{K}{2T(vol(V) + vol(E))^2}\left(\frac{vol(V)^2}{h_V\overline{K}^*_V} + \frac{vol(E)^2}{h_E\overline{K}^*_E}\right) \tag{6.21}$$

$$\geq \frac{K(vol(V)^2 + vol(E)^2)}{2MT(vol(V) + vol(E))^2} \geq \frac{K}{4MT}, \tag{6.22}$$

which ends the proof.

6.4.5 Buffer Plan Construction

After determining the set of selected intervals, we obtain the desired data placement strategy for the GPU buffer. However, to make this strategy operational during T-GNN training, we must construct concrete buffer plans that specify, for each batch,

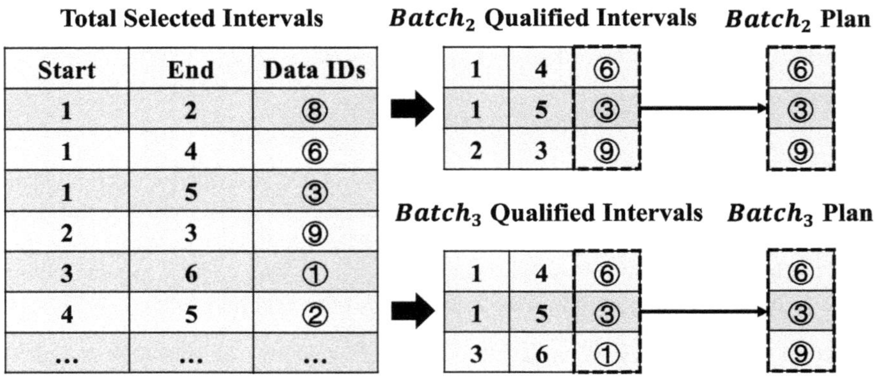

Fig. 6.12 Example of constructing the buffer plans out of selected intervals

which data IDs should be retained on the GPU. These plans serve as direct instructions for the GPU buffer to update its contents batch by batch.

To build these buffer plans, we need to search through the selected intervals, as illustrated in Fig. 6.12. For any given batch B_i, the corresponding buffer plan P_i includes all input data IDs associated with intervals $\sigma(\cdot)$ satisfying the following conditions:

$$s(\sigma(\cdot)) \leq i \tag{6.23}$$

$$e(\sigma(\cdot)) > i \tag{6.24}$$

where $s(\cdot)$ and $e(\cdot)$ denote the start and end batch indices of an interval, respectively.

A straightforward implementation involves iterating over all selected intervals for each batch and checking the above conditions. This results in a total time complexity of $\mathcal{O}(B \cdot n)$, where n is the number of selected intervals and B is the number of training batches. However, under large-scale settings typical in T-GNN training, n can be very large, making this naive approach computationally prohibitive. As demonstrated in the experiments later, the overhead of such brute-force plan construction can even exceed the total training time.

To address this inefficiency, we observe a critical temporal relationship: for any two consecutive batches, the qualified intervals for batch i are largely shared with those for batch $i - 1$. Specifically, all intervals with $s(\sigma(\cdot)) < i$ and $e(\sigma(\cdot)) > i$ (i.e., spanning batch i) must have also been valid for batch $i - 1$. Exploiting this overlap, we design an efficient construction method called ZigZag, which incrementally generates buffer plans using forward and backward scans:

- **Forward scan**: Adds newly qualified intervals starting at batch i.
- **Backward scan**: Retains previously qualified intervals that are still valid for batch i.

Algorithm 4 Buffer Plan Construction Policy ZigZag

Input: Selected interval list $\tilde{L} = [\sigma(1), \sigma(2), \ldots, \sigma(n)]$; number of batches B.
Output: Buffer update plan list P.
1: Initialize forward pointer $q \leftarrow 0$;
2: Initialize active index list $D \leftarrow [\ ]$;
3: Initialize output plan list $P \leftarrow [\ ]$;
4: Sort $\tilde{L}$ in ascending order of $s(\cdot)$;
5: **for** $i = 1$ to B **do**
6:     $\hat{D} \leftarrow [\ ]$
7:     **while** $\tilde{L}[q][0] \leq i$ **do** ▷ Forward scan
8:         $q \leftarrow q + 1$
9:         Append q to $\hat{D}$
10:     **end while**
11:     **for** d in D **do** ▷ Backward scan
12:         **if** $\tilde{L}[d][1] > i$ **then**
13:             Append d to $\hat{D}$
14:         **end if**
15:     **end for**
16:     $ID\_list \leftarrow$ Extract_ID$(\hat{D}, \tilde{L})$
17:     Append ID_list to P
18:     $D \leftarrow$ Update$(\hat{D}, i)$
19: **end for**

The complete procedure is shown in Algorithm 4. By maintaining continuity in the search space between successive batches, ZigZag significantly reduces redundant computation. Rather than scanning the entire interval list for each batch, it incrementally expands and filters a much smaller candidate set, enabling fast construction of per-batch buffer plans.

The detailed procedure of the buffer plan construction policy ZigZag, implemented by the plan generator in SIMPLE, is outlined in Algorithm 4. The algorithm begins by initializing the forward-search pointer q, the backward-search index list D, and an empty buffer plan list P (Lines 1–3). It then sorts the selected interval list $\tilde{L}$ in ascending order of their start batch indices (Line 4), which enables a clean separation of forward and backward search operations in the subsequent steps.

For each batch, ZigZag generates the corresponding buffer plan. It first initializes a temporary list $\hat{D}$ to store the indices of qualified intervals for the current batch (Line 6). The algorithm then performs a forward scan to collect newly starting intervals (Lines 7–10) and a backward scan to retain previously selected intervals that remain valid (Lines 11–15). Once all qualifying intervals are identified, ZigZag extracts the corresponding input data IDs (Line 16) and appends them to the buffer plan list P (Line 17). Finally, the candidate list D is updated to support backward scanning in the next iteration (Line 18).

Time Complexity. The ZigZag policy significantly reduces the computational cost of buffer plan construction by avoiding full scans over all candidate intervals for each batch. Its total time complexity is $\mathcal{O}(B \cdot \bar{l})$, where $\bar{l}$ denotes the average number of qualified intervals per batch. Since $\bar{l} \ll n$ (with n being the total number of candidate intervals), this method achieves substantial acceleration in practice.

6.4.6 Experiments

Experimental Setups

Datasets. Table 6.6 summarizes the key statistics of the datasets used in our experiments. We adopt four large-scale real-world dynamic graphs widely used in prior studies [15–17], each containing millions of temporal interactions. Specifically, *LastFM* [9] records user-song interaction events over the span of one month. *Wiki-Talk* [30] and *Stack-Overflow* [31] capture user-user interactions on their respective platforms. *GDELT* [17] is a large-scale temporal knowledge graph extracted from the GDELT 2.0 Event Database [19], comprising nearly one billion events.

For dataset partitioning, we follow a consistent chronological split for *LastFM*, *Wiki-Talk*, and *Stack-Overflow* as adopted in previous works [5, 6, 15], dividing each dataset into training (70%), validation (15%), and test (15%) sets. In the case of *GDELT*, we follow the same temporal split used in TGL [17], assigning interactions occurring before 2019, in 2019, and in 2020 to the training, validation, and test sets, respectively.

Baseline System and T-GNN Backbones. We compare SIMPLE against TGL [17], a state-of-the-art system designed for efficient and scalable T-GNN training. While DGL [27] is a popular framework for static graph learning, it currently lacks native support for T-GNN training and is therefore excluded from comparison. We also omit the original standalone implementations of T-GNN models, as TGL, equipped with a high-performance parallel temporal sampler and optimized data pipeline, consistently outperforms these original baselines.

For the backbone T-GNN architectures, we evaluate three representative models that capture distinct design philosophies: **(1)** *TGAT* [5], which encodes temporal information via random Fourier features and applies an attention mechanism similar to that in static GNNs; **(2)** *TGN* [6], which maintains evolving node state vectors to model temporal dependencies; and **(3)** *Orca* [16], a highly efficient model that lever-

Table 6.6 Summary of statistics of the dynamic graphs

Dataset	|V|	|E|	d_v	d_e	max(t)
LastFM	2 K	1.3 M	128	128	1.3e8
Wiki-Talk	1.1 M	7.8 M	172	172	2.0e8
Stack-Overflow	2.6 M	63.4 M	172	172	2.4e8
GDELT	17 K	191.3 M	413	186	1.8e5

|V| and |E| represent the number of nodes and edges. d_v and d_e denote the dimension of node features and edge features. max(t) shows the maximum edge timestamp

ages intermediate embedding reuse for computation reduction. Other models [7, 9, 15, 32] generally build on the foundational ideas of the aforementioned models, often by incorporating more complex architectures [15, 33]. Since the training workflows of these extended models are structurally similar to those of TGAT, TGN, and Orca, we exclude them from our evaluation for clarity and conciseness.

Downstream Task. In line with previous research [5–8, 15–17], our main downstream task is link prediction. The objective is to predict whether a future interaction will occur between a given pair of nodes. Since the original datasets only record observed (i.e., positive) interactions, we generate an equal number of negative samples (non-interacting node pairs) to construct a balanced evaluation setting. For performance evaluation, we adopt the Average Precision (AP) metric on the test set, following the standard protocol established in prior works [6, 8, 15–17].

It is important to note that we do not directly compare the predictive accuracy of T-GNN models trained using TGL and SIMPLE. This is because both systems serve as general-purpose training frameworks capable of efficiently training any T-GNN model without introducing additional components (e.g., data augmentation [38]) that might affect model effectiveness. The focus of our evaluation is on training efficiency and system design rather than on boosting raw model performance.

Configurations. To ensure a fair comparison between TGL and SIMPLE, we align all training configurations across both systems. We adopt the top-k recent neighbor sampling strategy [6] for all underlying T-GNN models, with k set to 10–matching the default setting used in TGL. Both systems utilize a standard 2-layer message passing scheme [5, 6, 16]. For batching, we set the batch size to 2000 for LastFM and Wiki-Talk, and 4000 for Stack-Overflow and GDELT. Regarding the GPU buffer in SIMPLE, we define its size as a user-configurable fraction of the total memory footprint of all input data. In our main experiments, this ratio is fixed at 0.1 by default. Other model-related hyperparameters remain unchanged from the original TGL implementation [17], as we do not perform additional tuning. Training is terminated once the validation performance fails to improve for three consecutive epochs. All experiments are conducted on a single machine equipped with one NVIDIA GeForce RTX 3090 GPU (24GB HBM), a 96-core Intel(R) Xeon(R) Gold 5220R CPU @ 2.20GHz, and 256GB of main memory. Reported results are averaged over three independent runs.

Training Time Comparison

Figure 6.13 presents the training time comparison between SIMPLE and TGL. Across all evaluated settings, SIMPLE achieves a 1.8× to 3.8× acceleration over TGL, with a speedup greater than 2× observed in 11 out of 12 cases. The primary factor contributing to this acceleration is the substantial reduction in CPU-GPU data loading overhead. As illustrated in Fig. 6.14, SIMPLE reduces this overhead by 80.5% to 96.5% through its dynamic data placement mechanism.

These results further demonstrate the versatility of SIMPLE in enhancing training efficiency across different T-GNN models. Notably, the extent of data loading cost

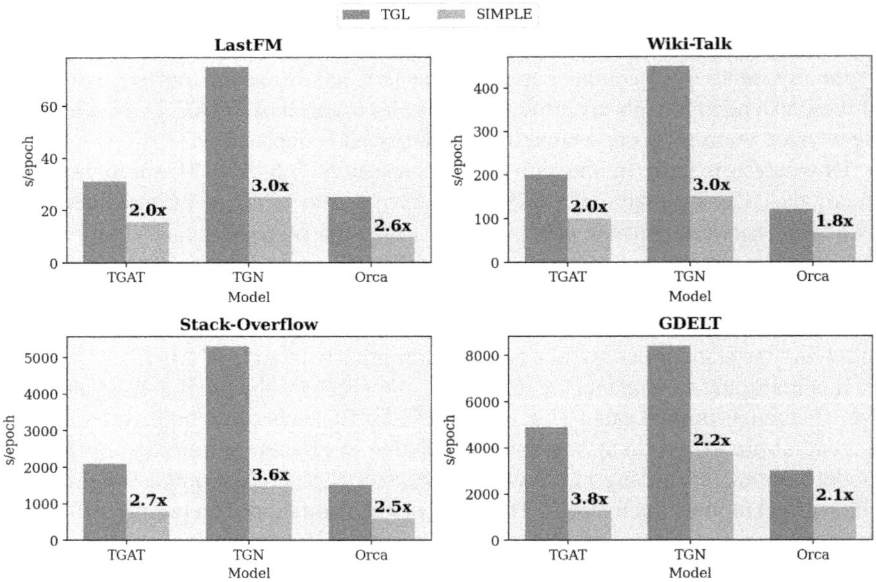

Fig. 6.13 Per-epoch training time(s) on LastFM, Wiki-Talk, Stack-Overflow and GDELT datasets for different T-GNN models under TGL and SIMPLE

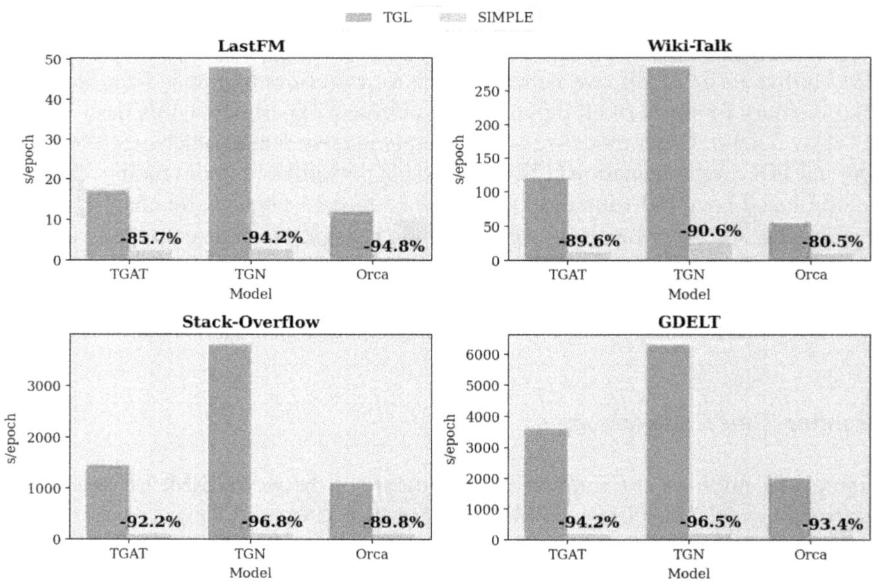

Fig. 6.14 Per-epoch data loading time(s) on LastFM, Wiki-Talk, Stack-Overflow, and GDELT datasets for different T-GNN models under TGL and SIMPLE

Table 6.7 The effect of buffer plan construction policy ZigZag by SIMPLE (unit:s)

Dataset	Model	w/o	w
LastFM	TGAT	1.65 (1.06×)	1.56
	TGN	1.70 (1.09×)	1.56
	Orca	1.56 (1.24×)	1.33
Wiki-Talk	TGAT	174.22 (17.44×)	9.99
	TGN	173.70 (16.88×)	10.29
	Orca	83.12 (9.69×)	8.58
Stack-Overflow	TGAT	10275.16 (18.48×)	556.08
	TGN	14496.90 (41.63×)	348.26
	Orca	5406.47 (19.53×)	276.84
GDELT	TGAT	10772.73 (306.57×)	35.14
	TGN	10111.27 (203.98×)	49.57
	Orca	6499.36 (193.61×)	33.57

'w' and 'w/o' respectively buffer plan construction with or without ZigZag

varies among the models. For instance, on the largest dataset GDELT, the data loading overhead for Orca is roughly one-third that of TGN. This is because Orca leverages intermediate embedding reuse to implicitly prune the temporal computation graph, thereby avoiding redundant input data fetches. Even under such optimizations, SIMPLE is able to further reduce Orca's data loading cost by 80.5% to 94.8%, and consistently achieves over 2× training speedup in most cases.

Evaluation of Additional Costs

We assess the additional overhead introduced by SIMPLE, specifically focusing on preprocessing, GPU buffer lookup, and GPU buffer update. As shown in Fig. 6.15, the preprocessing phase is lightweight, consuming only 6.6% of the total end-to-end training time on average. This efficiency is largely attributed to the ZigZag buffer plan construction strategy. As reported in Table 6.7, ZigZag accelerates buffer plan generation by up to two orders of magnitude, especially for large-scale datasets like GDELT. This acceleration stems from the significant reduction in interval search space.

Table 6.8 presents the runtime overhead of GPU buffer lookup and update. GPU buffer lookup introduces negligible cost—approximately 2% of the overall training time. GPU buffer update is also lightweight, accounting for only 2.0% to 23.0% of training time across different settings. These results demonstrate the effectiveness of the loader and updater modules implemented by SIMPLE.

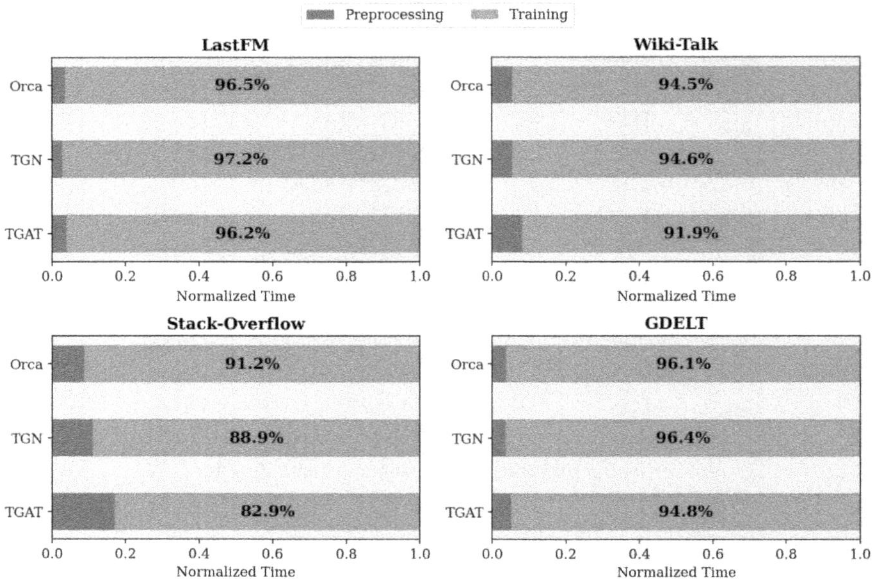

Fig. 6.15 Preprocessing and overall training time (excluding the preprocessing) comparison on LastFM, Wiki-Talk, Stack-Overflow, and GDELT datasets

Table 6.8 Per-epoch time of Buffer lookup and Buffer Update (unit:s)

Dataset	Model	Buffer Lookup	Buffer Update
LastFM	TGAT	0.4 (2.5%)	0.5 (3.1%)
	TGN	0.4 (1.6%)	0.5 (2.0%)
	Orca	0.2 (2.1%)	0.3 (3.1%)
Wiki-Talk	TGAT	2.0 (2.0%)	6.8 (6.8%)
	TGN	2.2 (1.4%)	6.9 (4.4%)
	Orca	1.2 (1.8%)	6.6 (10.1%)
Stack-Overflow	TGAT	13.6 (1.4%)	123.8 (13.0%)
	TGN	13.6 (1.0%)	110.7 (7.9%)
	Orca	6.3 (1.0%)	121.3 (18.6%)
GDELT	TGAT	31.7 (1.7%)	279.1 (15.1%)
	TGN	31.7 (1.3%)	285.8 (11.7%)
	Orca	24.1 (2.0%)	273.6 (23.0%)

Numbers inside the parenthesis indicate the percentage of the per-epoch training time

Impact of GPU Buffer Budget Ratio

We further conduct a case study on how varying the GPU buffer budget ratio affects system performance, using the TGN model on the Wiki-Talk and Stack-Overflow datasets. Results are shown in Table 6.9. As the buffer budget ratio increases, both data loading and overall training time decrease. Even with a minimal ratio of 1%, we observe substantial performance improvements—up to a 3.2× speedup. However, the marginal gains diminish as the ratio continues to grow, highlighting the effectiveness of SIMPLE's strategic dynamic data placement. In practice, allocating only 1–10% of GPU memory to the buffer already yields significant reductions in data loading cost.

Evaluation of Data Placement Strategy

We further implement two alternative data placement strategies within SIMPLE for comparison: a state-of-the-art static placement strategy and the classic Least Recently Used (LRU) policy [25]. The static strategy preloads a subset of input data that contributes the most to overall data loading volume, keeping them fixed in the GPU buffer throughout training. In contrast, LRU dynamically manages the buffer by evicting the Least Recently Used data after each batch.

To evaluate the effectiveness of SIMPLE's data placement strategy, we conduct experiments on Wiki-Talk and Stack-Overflow datasets using TGN as the backbone model, and set the GPU buffer budget ratio to 0.1 for all methods to ensure fairness. The results, summarized in Table 6.10, show that SIMPLE consistently achieves the lowest data loading cost and the fastest training time across both datasets.

Among the baselines, the LRU strategy outperforms the static strategy in terms of data loading efficiency, underscoring the benefit of dynamic buffer management

Table 6.9 Effect of different buffer budget ratios

Dataset	Ratio (%)	Data loading	Training
Wiki-Talk	0	289.3	462.2
	1	40.8	168.1
	5	28.0	153.7
	10	25.0	151.7
	20	20.2	146.2
Stack-Overflow	0	3571.5	5291.2
	1	470.4	1676.1
	5	290.3	1586.8
	10	113.8	1438.2
	20	112.4	1408.7

The per-epoch time (unit:s) is reported. 'Training' indicates the per-epoch training time excluding the preprocessing

Table 6.10 The effect of different data placement strategies. Per-epoch time (unit:s) is reported

Dataset	Strategy	Data loading	Training
Wiki-Talk	Static	58.4	184.1
	LRU	43.3	287.2
	SIMPLE's	25.0	151.7
Stack-Overflow	Static	1726.9	3319.6
	LRU	540.9	2614.3
	SIMPLE's	113.8	1438.2

'Training' indicates the per-epoch training time excluding the preprocessing

in T-GNN training. However, due to its reactive, per-batch decision-making based solely on recent usage, LRU can incur higher overall training costs compared to the static strategy. Moreover, LRU yields less significant reductions in data loading cost than SIMPLE, with the performance gap widening on the larger Stack-Overflow dataset.

This disparity arises because LRU relies only on limited historical access patterns, whereas SIMPLE generates buffer plans using global data occurrence statistics, offering a more informed and strategic approach. These findings highlight the importance of incorporating global data usage patterns when designing GPU buffer management policies for efficient T-GNN training.

6.5 Conclusion

Existing T-GNN training pipelines suffer from scalability issues due to ill-suited batching and high input data loading costs, which severely limit their efficiency on large-scale graphs. To address these two issues, we first propose ETC, a general-purpose framework designed to enable efficient T-GNN training on large-scale dynamic graphs. ETC introduces a novel batching strategy that supports large batch sizes while effectively mitigating the impact of information loss, thereby enhancing model computation efficiency. In addition, it features a three-step data access policy tailored for temporal neighbor retrieval, significantly reducing the overhead associated with input data access. To further improve efficiency, ETC employs a lightweight pipeline mechanism that decouples data access and model computation. Extensive experiments demonstrate that ETC achieves 1.6× to 62.4× training speedup over state-of-the-art T-GNN training frameworks across various models and large-scale datasets.

Then, we introduce SIMPLE, a general-purpose system designed to enable efficient T-GNN training on large-scale dynamic graphs. SIMPLE tackles the excessive CPU-GPU data loadingby employing a dynamic data placement strategy that significantly reduces data transfer overhead during training. Empirical evaluations demonstrate that SIMPLE lowers data loading costs by 80.5–96.8% and delivers an average 2.6× speedup in training time compared to the state-of-the-art system TGL. Looking ahead, it is promising to expand SIMPLE to support distributed training, addressing the increasing demand for scalability in industrial-scale dynamic-graph learning.

References

1. Bryan Perozzi, Rami Al-Rfou, Steven Skiena. 2014. Deepwalk: Online learning of social representations. In *KDD*. 701–710.
2. Aditya Grover, Jure Leskovec. 2016. node2vec: Scalable feature learning for networks. In *KDD*. 855–864.
3. Will Hamilton, Zhitao Ying, and Jure Leskovec. 2017a. Inductive representation learning on large graphs. *Advances in neural information processing systems* 30 (2017).
4. Hanqing Zeng, Hongkuan Zhou, Ajitesh Srivastava, Rajgopal Kannan, Viktor K. Prasanna. 2020. GraphSAINT: Graph sampling based inductive learning method. In *ICLR*.
5. Da Xu, Chuanwei Ruan, Evren K rpeoglu, Sushant Kumar, Kannan Achan. 2020. Inductive representation learning on temporal graphs. In *ICLR*.
6. Emanuele Rossi, Ben Chamberlain, Fabrizio Frasca, Davide Eynard, Federico Monti, Michael M. Bronstein. 2020. Temporal Graph Networks for Deep Learning on Dynamic Graphs. In *ICML 2020 Workshop on Graph Representation Learning*.
7. Yanbang Wang, Yen-Yu Chang, Yunyu Liu, Jure Leskovec, Pan Li. 2021. Inductive Representation Learning in Temporal Networks via Causal Anonymous Walks. In *ICLR*.
8. Xuhong Wang, Ding Lyu, Mengjian Li, Yang Xia, Qi Yang, Xinwen Wang, Xinguang Wang, Ping Cui, Yupu Yang, Bowen Sun, Zhenyu Guo. 2021. Apan: Asynchronous propagation attention network for real-time temporal graph embedding. In *Proceedings of the 2021 International Conference on Management of Data*. pp. 2628–2638.
9. Srijan Kumar, Xikun Zhang, Jure Leskovec. 2019. Predicting dynamic embedding trajectory in temporal interaction networks. In *SIGKDD*. pp. 1269–1278.
10. Thomas N. Kipf and Max Welling. 2017a. Semi-Supervised Classification with Graph Convolutional Networks. In *ICLR*. https://openreview.net/forum?id=SJU4ayYgl
11. Petar Velickovic, Guillem Cucurull, Arantxa Casanova, Adriana Romero, Pietro Li , Yoshua Bengio. 2018. Graph Attention Networks. In *ICLR*.
12. Palash Goyal, Sujit Rokka Chhetri, Arquimedes Canedo. 2020. dyngraph2vec: Capturing network dynamics using dynamic graph representation learning. *Knowledge-Based Systems*. 187.
13. Ehsan Hajiramezanali, Arman Hasanzadeh, Krishna R. Narayanan, Nick Duffield, Mingyuan Zhou, Xiaoning Qian. 2019. Variational graph recurrent neural networks. In *NIPS*. 10700–10710.
14. Aldo Pareja, Giacomo Domeniconi, Jie Chen, Tengfei Ma, Toyotaro Suzumura, Hiroki Kanezashi, Tim Kaler, Tao B. Schardl, Charles E. Leiserson. 2020. Evolvegcn: Evolving graph convolutional networks for dynamic graphs. In *AAAI*. 5363–5370.
15. Yiming Li, Yanyan Shen, Lei Chen, and Mingxuan Yuan. 2023b. Zebra: When Temporal Graph Neural Networks Meet Temporal Personalized PageRank. *VLDB* 16, 6 (2023), 1332–1345.
16. Yiming Li, Yanyan Shen, Lei Chen, and Mingxuan Yuan. 2023a. Orca: Scalable Temporal Graph Neural Network Training with Theoretical Guarantees. *Proc. ACM Manag. Data* 1, 1 (2023), 52:1–52:27. https://doi.org/10.1145/3588737
17. Hongkuan Zhou, Da Zheng, Israt Nisa, Vasileios Ioannidis, Xiang Song, George Karypis. 2022. TGL: A General Framework for Temporal GNN Training on Billion-Scale Graphs. *PVLDB*.
18. Rakshit Trivedi, Mehrdad Farajtabar, Prasenjeet Biswal, Hongyuan Zha. 2019. Dyrep: Learning representations over dynamic graphs. In *ICLR*.
19. Leetaru Kalev, Philip A. Schrodt. 2013. Gdelt: Global data on events, location, and tone, 1979–2012. In *ISA Annual Convention*. vol. 2, no. 4, 1–49.

20. Zhiqi Lin, Cheng Li, Youshan Miao, Yunxin Liu, and Yinlong Xu. 2020a. Pagraph: Scaling gnn training on large graphs via computation-aware caching. In *SoCC*. 401–415.
21. Jianbang Yang, Dahai Tang, Xiaoniu Song, Lei Wang, Qiang Yin, Rong Chen, Wenyuan Yu, and Jingren Zhou. 2022a. GNNLab: a factored system for sample-based GNN training over GPUs. In *EuroSys*. 417–434.
22. Xin Zhang, Yanyan Shen, Yingxia Shao, and Lei Chen. 2023a. DUCATI: A Dual-Cache Training System for Graph Neural Networks on Giant Graphs with the GPU. *Proc. ACM Manag. Data* 1, 2 (2023), 166:1–166:24. https://doi.org/10.1145/3589311
23. Rong Zhu, Kun Zhao, Hongxia Yang, Wei Lin, Chang Zhou, Baole Ai, Yong Li, Jingren Zhou. 2019. AliGraph: A Comprehensive Graph Neural Network Platform. *PVLDB*. 12(12): 2094-2105.
24. Seungwon Min, Kun Wu, Mert Hidayetoglu, Jinjun Xiong, Xiang Song, Wen-Mei Hwu. 2022. Graph neural network training and data tiering. In *KDD*. 3555–3565.
25. Elizabeth J. O'Neil, Patrick E. O'Neil, Gerhard Weikum. 1993. The lru-k page replacement algorithm for database disk buffering. In *SIGMOD*. 297–306.
26. Bernhard Korte, Jens Vygen. 2008. Combinatorial optimization: Theory and algorithms. *Springer*.
27. Minjie Wang, Da Zheng, Zihao Ye, Quan Gan, Mufei Li, Xiang Song, Jinjing Zhou, Chao Ma, Lingfan Yu, Yu Gai, et al. 2019a. Deep graph library: A graph-centric, highly-performant package for graph neural networks. *arXiv preprint* arXiv:1909.01315 (2019).
28. Adam Paszke, Sam Gross, Francisco Massa, Adam Lerer, James Bradbury, Gregory Chanan, Trevor Killeen, Zeming Lin, Natalia Gimelshein, Luca Antiga, Alban Desmaison, Andreas K pf, Edward Z. Yang, Zachary DeVito, Martin Raison, Alykhan Tejani, Sasank Chilamkurthy, Benoit Steiner, Lu Fang, Junjie Bai, Soumith Chintala. 2019. Pytorch: An imperative style, high-performance deep learning library. In *NeurIPS*. 8024–8035.
29. Charles R. Harris, K. Jarrod Millman, St fan van der Walt, Ralf Gommers, Pauli Virtanen, David Cournapeau, Eric Wieser, Julian Taylor, Sebastian Berg, Nathaniel J. Smith, Robert Kern, Matti Picus, Stephan Hoyer, Marten H. van Kerkwijk, Matthew Brett, Allan Haldane, Jaime Fern ndez del R o, Mark Wiebe, Pearu Peterson, Pierre G rard-Marchant, Kevin Sheppard, Tyler Reddy, Warren Weckesser, Hameer Abbasi, Christoph Gohlke, Travis E. Oliphant. 2020. Array programming with numpy. *Nature*. 585: 357–362.
30. "Wiki-talk," [2023]. [Online]. Available: http://snap.stanford.edu/data/wiki-talk-temporal.html
31. "Stack-overflow," [2023]. [Online]. Available: https://snap.stanford.edu/data/sx-stackoverflow.html
32. Ming Jin, Yuan-Fang Li, Shirui Pan. 2022. Neural temporal walks: Motif-aware representation learning on continuous-time dynamic graphs. In *NeurIPS*.
33. Weilin Cong, Si Zhang, Jian Kang, Baichuan Yuan, Hao Wu, Xin Zhou, Hanghang Tong, Mehrdad Mahdavi. 2023. Do we really need complicated model architectures for temporal networks? In *ICLR*.
34. "The technical report of etc," [2023]. [Online]. Available: https://github.com/eddiegaoo/ETC/blob/main/Technical-report.pdf
35. Silvano Martello, Paolo Toth. 1990. Knapsack problems: algorithms and computer implementations. *John Wiley & Sons, Inc.*.
36. Arnaud Fréville. 2004. The multidimensional 0–1 knapsack problem: An overview. *European Journal of Operational Research*. 155(1): 1–21.
37. Valentina Cacchiani, Manuel Iori, Alberto Locatelli, Silvano Martello. 2022. Knapsack problems–an overview of recent advances. part ii: Multiple, multidimensional, and quadratic knapsack problems. *Computers & Operations Research*. 143: 105693.
38. Guoliang Li, Xuanhe Zhou, Lei Cao. 2021. Ai meets database: Ai4db and db4ai. In *SIGMOD*. 2859-2866. NeutronOrch: Rethinking Sample-based GNN Training under CPU-GPU Heterogeneous Environments. *Proc. VLDB Endow.* 17, 8 (2024), 1995–2008. https://doi.org/10.14778/3659437.3659453